MULTIVERSE THEORIES

If the laws of nature are fine-tuned for life, can we infer other universes with different laws? How could we even test such a theory without empirical access to those distant places? Can we believe in the multiverse of the Everett interpretation of quantum theory or in the reality of other possible worlds as advocated by philosopher David Lewis? At the intersection of physics and philosophy of science, this book outlines the philosophical challenge to theoretical physics in a measured, well-grounded manner. The origin of multiverse theories is explored within the context of the fine-tuning problem, and a systematic comparison between the various different multiverse models is included. Cosmologists, high-energy physicists, and philosophers including graduate students and researchers will find a systematic exploration of such questions in this important book.

SIMON FRIEDERICH is Associate Professor of Philosophy of Science at the University of Groningen and holds PhDs in both physics and philosophy. His work on multiverse theories was supported by a Veni grant from the Netherlands Organization for Scientific Research.

MULTIVERSE THEORIES

A Philosophical Perspective

SIMON FRIEDERICH

University of Groningen

CAMBRIDGE
UNIVERSITY PRESS

University Printing House, Cambridge CB2 8BS, United Kingdom

One Liberty Plaza, 20th Floor, New York, NY 10006, USA

477 Williamstown Road, Port Melbourne, VIC 3207, Australia

314–321, 3rd Floor, Plot 3, Splendor Forum, Jasola District Centre, New Delhi – 110025, India

79 Anson Road, #06–04/06, Singapore 079906

Cambridge University Press is part of the University of Cambridge.

It furthers the University's mission by disseminating knowledge in the pursuit of education, learning, and research at the highest international levels of excellence.

www.cambridge.org
Information on this title: www.cambridge.org/9781108487122
DOI: 10.1017/9781108765947

© Simon Friederich 2021

First published 2021

Printed in the United Kingdom by TJ Books Limited, Padstow Cornwall

A catalogue record for this publication is available from the British Library.

Library of Congress Cataloging-in-Publication Data
Names: Friederich, Simon, author.
Title: Multiverse theories : a philosophical perspective / Simon Friederich.
Description: First edition. | New York : Cambridge University Press, 2021.
| Includes bibliographical references and index.
Identifiers: LCCN 2020016490 (print) | LCCN 2020016491 (ebook) |
ISBN 9781108487122 (hardback) | ISBN 9781108765947 (epub)
Subjects: LCSH: Quantum theory. | Science–Philosophy.
Classification: LCC QC174.12 .F753 2021 (print) | LCC QC174.12 (ebook) |
DDC 530.12–dc23
LC record available at https://lccn.loc.gov/2020016490
LC ebook record available at https://lccn.loc.gov/2020016491

ISBN 978-1-108-48712-2 Hardback

To my children, who have entered my life since I have been
working on this book

Contents

10   Other Multiverses   155
     10.1 The Everettian Quantum Multiverse   155
     10.2 Modal Realism   168
     10.3 A Multiverse of Mathematical Structures?   173

11   Outlook   178
     11.1 Is the Multiverse Even Science?   178
     11.2 Whither Physics?   180

     References   185
     Author Index   197
     Subject Index   199

Preface

The history of this book starts in 2008, with a visit that my wife (then girlfriend) and I, at that time living in Heidelberg, received from our friend Marlene Weiß, former string theorist, recently turned journalist. Marlene told us about a radio feature she was making on a topic at the intersection of physics and philosophy of science: the "multiverse." Despite being an aspiring philosopher of physics, I had never heard of this "philosophical challenge to theoretical physics," as Marlene's radio feature puts it. She was in the process of interviewing numerous expert physicists and philosophers for it – among them my PhD supervisor in physics, Christof Wetterich, and my (nowadays) colleague Richard Dawid, string theorist turned philosopher of science.[1]

I found what Marlene told us about the multiverse debate very interesting, and I felt a bit ashamed for not yet being familiar with it. I also made a mental note to read and think more about it when I got a chance. A few months later – after Marlene had already abandoned the topic – I took the time to do so and started reading about the multiverse debate. In doing so, I began to understand what the anthropic principles are about, and I also began to think that the intriguing puzzles of self-locating belief, which seemed, in some not quite clear way, connected to the anthropic principles.

Later, during the two years in which I worked at the University of Göttingen (2012–2014), I had the opportunity to teach several courses on these topics, among them colloquia for more advanced students, in which I could fruitfully discuss many of the issues treated here and develop my thoughts on them. In 2016, now working at the University of Groningen in the Netherlands, I was lucky to obtain a Veni grant (NWO project number 275-20-065) from the Netherlands Organization

---

[1] A transcript of Marlene's highly recommendable feature (in German) is online and can be found at www .deutschlandfunkkultur.de/manuskript-die-beste-aller-moglichen-welten-pdf.media .d3645e999b4cac85489d482043313616.pdf.

for Scientific Research (NWO) for a project on the "Epistemology of the Multiverse," which gave me time and greatly facilitated my work on this book. Another facilitating factor has been my contact with the Munich Center for Mathematical Philosophy, notably the group of Stephan Hartmann. He was also my mentor in a habilitation process that resulted in an earlier version of this book.

Academics often have strong feelings about the views that they defend or criticize. This has its advantages and disadvantages. An advantage is that academics are often motivated by the passion they feel about their topics of interest to engage in great efforts in order to provide argumentative or empirical support for their conjectures and favorite hypotheses. Another advantage is that it makes them inventive in devising intelligent criticisms of hypotheses that they do not like. A disadvantage is that their passion tends to make their judgment biased, especially when it comes to assessing views to which they have already committed themselves publicly.

The considerations on multiverse theories in this book are somewhat unusual in this respect, for they have been developed and compiled by someone who has never had any strong feelings about their central topic; the question of whether there are other universes where physical parameters differ from those in our own leaves me rather cold, and so does the question of whether physical theories that entail the existence of other universes are empirically testable or not.

In analogy to the advantages and disadvantages of passion in academia just mentioned, my indifference about the truth and testability of multiverse theories has its corresponding disadvantages and advantages. A disadvantage has been that I have sometimes not been as motivated as I would have liked to have been to work on the topics treated in this book and sometimes pursued that work mostly for the sake of the intrinsically rewarding intellectual exercise. But a complementary advantage might be that the considerations presented in this book are developed from the relatively disinterested perspective of someone without any "skin in the game." My hope is that this will have allowed the present book to steer clear both of uncritical enthusiasm about multiverse ideology and of unconstructive bashing as unscientific of any serious mention of the possibility that there might be other universes.

Some of the thoughts and passages that appear in this book have appeared, sometimes in different or more preliminary form, in other publications by me. Those are the following:

- S. Friederich, 2018, Fine-tuning, *Stanford Encyclopedia of Philosophy*, ed. by E. N. Zalta, Fall 2018 edition.
- S. Friederich, 2017, Choosing beauty, *Logique et Analyse*, 60:449–463.
- S. Friederich, 2017, Resolving the observer reference class problem in cosmology, *Physical Review D*, 95:123520.

- S. Friederich, 2017, Fine-tuning as old evidence, double-counting, and the multiverse, *International Studies in the Philosophy of Science*, 31:363–377.
- S. Friederich, 2019, Reconsidering the inverse gambler's fallacy charge against the fine-tuning argument for the multiverse, *Journal for General Philosophy of Science*, 50:29-41.
- S. Friederich, 2019, A new fine-tuning argument for the multiverse, *Foundations of Physics*, 49:1011–1021.

Section 10.1.3, about the Sebens/Carroll approach to the probability problem of the Everett interpetation, has partly been shaped by discussions with Richard Dawid. Complementary to the considerations developed in that section, we provide further criticisms of the Sebens/Carroll approach in a forthcoming paper "Epistemic separability and Everettian branches – a critique of Sebens and Carroll," which is currently under review with a journal.

Among those who helped this book come to fruition, I would like to thank first of all Stephan Hartmann for hospitality in Munich and many lively discussions there. Thankfully, I received a plethora of stimulating comments and criticisms on an earlier incarnation of this book, which was accepted as a habilitation thesis at the University of Munich, by George Ellis, Dieter Lüst, John Norton, and Chris Smeenk. I would also like to thank Chris for inviting me to be a visiting scholar at the Rotman Institute of Philosophy at the University of Western Ontario in June 2019 and the members of the Rotman Institute for their amazing hospitality and a stimulating experience.

I am also grateful for useful comments on drafts of papers that morphed into parts of this book as well as for other rewarding exchanges with Guus Avis, Luke Barnes, Christoph Behrens, Jeremy Butterfield, Eric Curiel, Richard Dawid, Sean Gryb, Robert Harlander Richard Healey, Leah Henderson, Mike Hicks, Klaas Landsman, Casey McCoy, Felix Mühlhölzer, Jan-Willem Romeijn, Karim Thébault, Sylvia Wenmackers, and Claire Zukowski. I am also grateful to the anonymous referees who reviewed my papers for various journals and to the anonymous referees who were consulted by Cambridge University Press.

Moreover, I would like to thank members of the audience at seminars, conferences, and workshops in Brisbane, Bristol, Crete, Düsseldorf, Göttingen, Groningen, Hamburg, Munich, Sydney, Turku, Utrecht, and Wuppertal. Among the audience members whose questions I found particularly stimulating were David Albert, Harvey Brown, Pete Evans, Martin Gustafsson, Joel Katzav, James Ladyman, Marco Livio, Tim Maudlin, Dean Rickles, Simon Saunders, and David Wallace. For fantastic support in designing my grant application to the NWO, I am grateful to Jeanne Peijnenburg and David Atkinson.

Most of all, I would like to thank my wife Andrea and our five daughters for making my life as happy and rewarding as it thankfully is.

Part I

Setting the Stage

1

Introduction

1.1 Multiverse Theories and Why to Consider Them

Multiverse theories are physical theories according to which we have empirical access only to a tiny part of reality that may not at all be representative of the whole. According to such theories, the laws of nature are *environmental* in the sense that other parts of reality to which we may not have any causal and empirical access have very different laws – or that there are at least certain aspects of the laws of nature that are very different in those other parts of reality.

Multiverse theories differ on what those other "parts of reality" are. They can, for example, be distant space-time regions that are so far from us that we cannot causally interact with any objects located there. Or they can be distinct "subuniverses" of an overarching collection of separate universes – a "multiverse" perhaps more in the original sense of the word – which have different laws of nature and may not even stand in any spatiotemporal relations to each other. For the purposes of this book, I refer to all types of physical theories according to which reality is in some sense much larger and more diverse than what we have access to as "multiverse theories."

This characterization of multiverse theories is clearly rough and imprecise. But it suffices to make it plausible that theories qualifying as "multiverse theories" in my sense are likely to be interesting from a philosophical point of view. Indeed, they give rise to intriguing epistemological challenges.

To begin with, it seems hard to deny the possibility in principle that a multiverse theory might hold and that the laws of nature in our "universe" (whatever exactly qualifies as such) are environmental in that they may not be representative of the laws across all the many constituent "universes" of the overall multiverse. But since those hypothetical other universes are, by assumption, causally inaccessible to us, we cannot convince ourselves of their existence directly through observations and cannot check this key aspect of those theories empirically. The best we can hope for is to identify aspects of those theories that make them testable by means of observations confined to our own universe and – if those tests are successful – to indirectly infer the existence and properties of the other universes entailed by them with more or less confidence. In this book, I investigate to what degree that hope to make multiverse theories susceptible to such indirect testing is actually realistic.

Why would we possibly want to consider multiverse theories at all if their very testability raises so complicated questions? One influential motivation to consider them is that several aspects of the laws of nature in our universe seem *fine-tuned for life*. Notably, this seems to hold for various features of the form of those laws themselves, for several constants that appear in those laws, and for the global boundary conditions of our universe that characterize its early stages. According to many physicists, had those features of the laws, constants, and boundary conditions been slightly different, life could probably not have existed in our universe, and so we could not have existed in it. In the eyes of many, the fact that we exist despite the fine-tuning of all those parameters cries out for an explanation. The truth of some multiverse theory may provide one.

The core idea of the suggested multiverse explanation of life's existence despite the required fine-tuning is that, if there is a sufficiently diverse multiverse where the parameters (describing the forms of the laws, the constants, and the boundary conditions) differ between universes, it is only to be expected that there are at least some universes where the parameters are right for life. As living organisms, we could not possibly have found ourselves in a universe that fails to be life friendly. This suggests that, under the assumption that there is a sufficiently diverse multiverse, it is neither surprising that there is at least one universe that is hospitable to life nor – since we could not have found ourselves in a life-hostile universe – that we find ourselves in a life-friendly one. Thus, our existence as forms of life, which seems baffling in view of the fine-tuned parameters that are needed for it, no longer seems surprising if we assume that our universe is actually part of a much larger multiverse with diverse environmental parameters.

This suggested inference to the existence of a multiverse as providing the best account of why there is life despite the required fine-tuning will be called the "standard fine-tuning argument for the multiverse" in what follows. I discuss it in detail in later chapters of this book.

But what concrete type of physical multiverse theory might provide us with a multiverse in the sense of the standard fine-tuning argument for a multiverse?

1.2 Types of Multiverse Theories

The simplest type of multiverse theory that could function in the standard fine-tuning argument for the multiverse is one that hypothesizes only a single, connected space-time manifold where certain constants – e.g., Newton's constant – vary over large temporal and/or spatial length scales. If the variation of the constants occurs on time or length scales that are astronomical but that can still be probed by us, this type of theory may not qualify, strictly speaking, as a "multiverse theory" in the present sense. But if the constants that vary across space or time according to it require fine-tuning to be compatible with life, it may nevertheless effectively play the role of a multiverse theory in the standard fine-tuning argument for the multiverse. Inasmuch as such theories are indeed empirically testable, the available evidence does not seem to provide significant support for them [Uzan, 2003].

Another type of multiverse theory that is straightforward to characterize is one according to which there is an ensemble of (real) spatiotemporally unconnected universes, all with laws of the same form as those in our universe but with different values of certain constants. Since the most established theories of modern fundamental[1] physics are the Standard Model of elementary particle physics (combining the electroweak theory and quantum chromodynamics) and general relativity, in such a multiverse, the universes would all be described by those theories, but with masses of elementary particles and interaction constants different in the different universes.

A drawback of this type of multiverse theory is that it has little to no independent motivation over and above the fine-tuning considerations. In contrast, the so-called *landscape multiverse* [Susskind, 2005], which results from combining string theory with certain models of inflationary cosmology, is an independently motivated cosmological scenario. As we will see in what follows, it can make a good claim to count as a multiverse theory in the sense of the standard fine-tuning argument for the multiverse.

1.2.1 Inflationary Cosmology

Inflationary cosmology, originally developed by Guth [2000], is currently the dominant theoretical framework of early-universe cosmology. It states that the very early universe expands (near-) exponentially fast, cooling down by many orders of magnitude, before transitioning to a period of much slower expansion and "reheating." The original motivation for inflationary cosmology was that it promised an explanation of otherwise puzzling cosmic coincidences – namely, the so-called flatness, horizon, and magnetic monopole problems of cosmology [Guth, 1981; Linde, 1982]; see [Guth, 2000] for a review. To appreciate the appeal

---

[1] Almost always, when I use the adjective "fundamental" in this book, it is meant in a loose sense, signifying something like "concerning the most basic entities and interactions that we have knowledge of." Except in the book's last chapter, I never use "fundamental" in the more ambitious sense in which one can reasonably ask whether there is an ultimate, fundamental, physical level where the edifice of physical theories "bottoms out."

of inflationary cosmology, it is worth briefly reviewing these problems. (Readers familiar with inflationary cosmology can skip this subsection.)

The flatness problem arises from the observation that the universe today is completely flat (it has zero curvature within the precision of our measurements) on large length scales. This is puzzling because, according to the well-understood dynamics governing the expansion of our universe in the past billions of years, any slight deviation from perfect flatness would have dramatically increased over time. This means that our universe must have been very flat indeed in its very early stages; i.e., it must have started out in some highly nongeneric, "fine-tuned" state of near-perfect flatness.

Inflationary cosmology supposedly solves this problem by resulting in a state with (very near-) zero curvature at its end, independently of how curvature was at its start. The inflationary expansion period, in other words, produces a universe that is so flat that the slower expansion process since inflation, which tended to increase any remaining curvature, has so far not resulted in any measurable deviation from it on large length scales.

The claim that inflation thereby solves the flatness problem is controversial. For example, Hollands and Wald [2002] criticize it by arguing that the universe must occupy a very specific kind of state in order to be at the onset of curvature-erasing inflation. According to this criticism, inflation merely substitutes one "fine-tuning" problem for another and, thus, does not really mean progress with respect to the flatness problem. (The general structure of fine-tuning problems will be discussed in Chapter 2.)

The horizon problem, in turn, arises from the fact that, again on very large length scales, the universe today seems almost completely homogeneous and isotropic. This is puzzling because distant regions that we now observe as having identical large-scale properties have never been in causal contact with each other – at least not if we extrapolate the known (noninflationary) dynamics of the expansion of our universe into the past. But if certain regions of the universe have never been in causal contact with each other, their homogeneity cannot be the result of a joint equilibration process. This makes their homogeneity and isotropy on large length scales at least prima facie very surprising.

Inflationary cosmology supposedly solves this problem by providing a mechanism of how distant regions with identical large-scale properties have been in causal contact after all: if there has been a very early inflationary period, the distant regions were once in causal contact after all, and their observed homogeneity and isotropy raise no great puzzles.

This suggested solution is not without its critics either. Hollands and Wald [2002] raise worries about it that parallel those that they have about inflation's suggested solution to the flatness problem.

Finally, the magnetic-monopole problem, arises if one assumes that a so-called *grand unified theory* (GUT) obtains, which entails the existence of stable magnetic monopoles. The motivation for such a theory is that it can, in principle, provide an elegant unification of the electroweak theory and quantum chromodynamics similarly to how the electroweak theory itself provids a unified account of electromagnetism and the weak nuclear interaction.

If magnetic monopoles are permitted by the laws of nature, one would expect them to be produced in abundance in the hot very early universe, and their absence from observation is thus puzzling. Inflationary cosmology would, in that case, provide an explanation of that absence because inflation could easily have diluted magnetic monopoles to the point of making them undetectable. The power of this argument for inflationary cosmology depends on how strong one takes the theoretical case for magnetic monopoles based on GUTs to be. In the contemporary theoretical environment, where considerations in favor of GUTs may seem less compelling than in the early 1980s, the argument for inflation based on magnetic monopole abundance may not be regarded as very strong.

As already indicated, it is somewhat controversial whether inflationary cosmology really solves the problems just outlined, which it was originally designed to solve. The question of whether it does so is related to the question of whether conditions that give rise to inflation are rather generic or, in fact, so specific that the challenge to account for why they might have been met seems as large as the explanatory challenge that inflation purportedly helps to address.

As pointed out by Hawking and Page [1988] and elaborated more recently by Shiffrin and Wald [2012], the phase space of general relativity is non-compact. Probabilities over entire space-time histories can only be defined if ambiguities are removed by choosing a regularization procedure. Because of the differences between viable regularization procedures, different accounts of the probability for inflation to happen – e.g., the conflicting ones given in Gibbons et al. [1987] and [Gibbons and Turok, 2008] – come to radically different conclusions regarding how "probable" inflation really is. Correspondingly, they differ on how much postulating an inflationary period can contribute to resolve the horizon and flatness problems.

Nowadays, it is no longer inflationary cosmology's potential to solve the horizon and flatness problems that is widely regarded as its most important attraction. Rather, its ability to make precise and accurate predictions concerning the spectrum of the cosmic microwave background (CMB) fluctuations is now seen as its most important empirical achievement. These fluctuations have recently measured with unprecedented accuracy by the Planck satellite [Planck Collaboration, 2016]. Overall, the observed fluctuation pattern corresponds very well with the predictions derived on the basis of at least some inflationary models [Martin, in press].

As observed by McCoy [2015], it is remarkable that the theory now apparently fares quite well from an empirical point of view even though its original motivation – that it allegedly solves the flatness, horizon, and magnetic monopole problems – is now no longer widely viewed as compelling.

Indeed, there is also some debate on how compelling the support really is that inflationary cosmology derives from its successful prediction of the observed CMB fluctuations pattern. Notably, it has been argued that certain noninflationary models of cyclic cosmology are just as good in predicting that pattern [Lehners and Steinhardt, 2013]. But the majority view seems to be that at least some inflationary models are superior in this respect [Linde, 2014].

If there really was a period of rapid inflation in the very early universe, what might have been the mechanism that drove it? According to most models of inflation, one or more scalar fields, the so-called inflaton(s), are the most likely culprits.

There has been some debate on whether the Higgs boson, which is responsible for the masses of several particles in the Standard Model of elementary particles, might be the inflaton. But in most models of inflation, the inflaton field is distinct from any known particle and only identified by its role in generating an inflationary period. In other words, in most models of inflation, as driven by an inflaton, the inflaton field must be postulated to fulfill precisely that purpose and has no independent motivation.

The predictive and explanatory successes of inflationary cosmology – which, as just outlined, may come with certain caveats – provide one of the main reaons for taking multiverse theories seriously. The reason is that, according to many inflaton models, notably ones in which the potential of the inflaton field depends quadratically on the field strength, island universe formation is globally "eternal." When it comes to an end, it does so only locally, resulting in the formation of a causally isolated space-time region that effectively behaves as an "island universe." This process of continuing island universe formation never stops. As a result of it, a vast (and, according to most models, infinite) "multiverse" of island universes is continually being produced [Guth, 1981].

Inflationary cosmology as a general framework should not be equated with eternal inflation. Notably, there are empirically viable inflaton models according to which the inflationary period globally does come to an end [Mukhanov, 2015], [Martin, in press, Sect. 7C]. As we will soon see, though, the idea of inflation being eternal gets further support and attraction when one adds string theory to the picture. Doing so also brings into play a natural way in which the laws of nature might be effectively different in the different island universes, yielding an actual multiverse scenario.

1.2.2 String Theory

String theory is one of the leading approaches – perhaps still *the* leading approach – to unify our best current theories of particle physics as collected in the Standard Model of elementary particle physics and Einstein's theory of general relativity. The objects that the theory posits are one-dimensional objects called "strings" and various higher-dimensional analogs commonly referred to as "branes." Particles that are familiar from elementary particle physics are recovered as excitation modes of strings as they appear to an observer who lacks an apparatus with the resolution required to resolve the string structure.

In order to have the potential to be empirically viable, string theory must be considered in a version that includes *supersymmetry*. According to the idea of supersymmetry, the two main types of particles, fermions and bosons, are connected by a symmetry operation in the mathematical sense – "supersymmetry" – that can be regarded as a generalization of the familiar space-time symmetries such as invariance of the laws under spatial rotations. If supersymmetry holds, each fermionic particle has a bosonic counterpart with otherwise very similar properties, and vice versa. However, no supersymmetric partners of particles known to exist have been found in any collider experiments yet: there is not a single fermion or boson, for which a candidate partner particle has been detected. It follows that the partner particles, if they exist, must have considerably higher masses than the known particles. This means that supersymmetry must be *broken* by some hitherto unknown mechanism that makes it undectable at so far accessible energy scales.

There is an independent line of reasoning in favor of supersymmetry, based on the concept of *naturalness*, which is reviewed in Section 2.2.2. Mainly based on the idea that the fundamental physical theories should be "natural" in the somewhat technical sense to be elucidated there, it was widely expected until some years ago that supersymmetric partner particles would soon be found in collider experiments. But this has not happened, and the failure to discover any direct evidence in favor of supersymmetry is now more and more widely seen as pointing to shortcomings of the naturalness criterion and, more specifically, a blow to the attractiveness of string theory, whose viability depends on supersymmetry being realized.

One of the most important arguments in favor of string theory is the *no alternatives argument*, formally developed by Dawid et al. [2015] and spelled out in detail in Dawid [2013]. Beyond motivating string theory as a potential unification of elementary particle physics and gravity, it observes that there are few, if any, serious *alternative* theories that offer the same potential for unification while being empirically adequate, and it concludes that this provides at least some degree of support for string theory. The no alternatives argument remains controversial, however, in particular, because it is doubtful whether we can ever have a sufficient overview of the space of theoretical possibilities, including hypothetical alternatives to string theory, to make such strong conclusions.

Another argument for string theory refers to the unexpected coherence of different theoretical paths to it that were originally regarded as independent of each other. Several ostensibly different and competing string theories were pursued until 1995. At that time, it became clear that these theories are connected by so-called *dualities*, which means that they can be mapped onto each others in a way that reveals their physical equivalence. Another important duality discovery is that of *Anti–de Sitter/conformal field theory* (AdS/CFT) duality. Exploiting this duality helps make the physical consequences of string theory more transparent, and it has found widespread applications in physics far beyond string theory.

String theory has some specific physical consequences, which are, in principle, empirically testable: notably, it entails "stringy" features of reality, which would become empirically manifest at very high energies close to the Planck scale (about 13 orders of magnitude larger than energies accessible at present-day colliders). The familiar phenomenology of "particles" in present-day high-energy physics is only an "effective" low-energy phenomenon from the string theoretic perspective.

Another consequence of string theory is that space-time has to be 10-dimensional for its supersymmetric version to be compatible with massive particles. Since space-time is manifestly *not* 10-dimensional at the level of our experiences, one must assume that six of the nine spatial dimensions are effectively "compactified" at short spatial length scales. From a theoretical point of view, this is entirely conceivable. So-called *Calabi-Yau manifolds* offer a variety of ways in which the spatial extra dimensions entailed by string theory might in principle be compactified.

String theory is now believed to harbor an enormous amount of lowest-energy states, so-called *vacua*. Already in the 1980s, the number of such vacua was found to be very large [Lerche et al., 1987], and it has since been estimated to be of an order of magnitude comparable to 10^{500} [Bousso and Polchinski, 2000]. At the level of human-scale observations and experiments, the specific properties of these different vacua would manifest themselves in terms of different parameters – i.e., different higher level physical laws and different values of the constants. That there are string theory vacua with small positive cosmological constants, as actually observed, was argued by Kachru et al. [2003] and seems now widely accepted.[2]

The plurality of effective low-energy laws to which string theory gives rise makes it very difficult to extract concrete empirical consequences from the theory.

---

[2] I would like to thank George Ellis for alerting me of the Kachru et al. [2003] paper and for sharing his critical perspective on the viability of the mechanism it suggests.

This makes string theory hard to test, and this has led some researchers to speak of a methodological crisis in fundamental physics [Smolin, 2006; Woit, 2006]. That string theory is now often considered in a multiverse setting, combined with eternal inflation, does not appease those critics, on the contrary.

1.2.3 The Landscape Multiverse

Eternal inflation and string theory are independent of each other: it may well be the case that one of those two theoretical ideas is realized while the other is not.

However, *if* one assumes that string theory holds in combination with some scenario of inflationary cosmology, then it seems natural to expect that the inflationary period will be eternal. At least somewhere, a metastable inflating state may initially be realized in the inflating cosmos that happens to decay into noninflating states forming island universes at decay rates that are smaller than the inflating state's own expansion rate. If that is the case, the expansion of the metastable inflating state globally never stops despite the ongoing "bubble formation" of island universes, which in turn continues indefinitely.

A cosmological setting in which string theory holds in combination with inflation being eternal may potentially give us a concrete instantiation of the general multiverse idea as outlined earlier. For if there are indeed infinitely many island universes, as entailed by eternal inflation, then all the different string theory vacua – corresponding to different higher-level physical laws and constants – might actually be realized in them. To make this scenario credible, a physical mechanism would be needed, which accounts for why and how different string theory vacua would be realized in the different island universes. If some such mechanism indeed exists and, as is widely believed, this landscape multiverse includes a universe with the same higher-level laws and constants as our own, it is a candidate multiverse scenario in the sense of the argument for a multiverse from fine-tuning for life.

With the combination of eternal inflation and string theory in form of the landscape multiverse, we have a concrete multiverse scenario with independently motivated pillars – i.e., a concrete candidate multiverse "theory." This underlines the pressing need to obtain a clearer perspective on the empirical testability of such theories. That need appears even more urgent in view of the fact that it seems doubtful whether the independent empirical motivation of inflationary cosmology through the CMB data and possibly the response to the flatness and horizon problems survive the shift to a multiverse setting. Ijjas et al. [2013] argue that the independent empirical motivation of inflationary cosmology, which they do not regard as compelling in view of the data from the Planck satellite (see Planck Collaboration [2016] for the most recent edition) anyway, breaks down completely when the shift to a multiverse setting is made. In Ijjas et al. [2017], they outline the gist of their worry about inflation being eternal as follows:

The worrisome implication [of eternal inflation] is that the cosmological properties of each patch [i.e., island universe] differ because of the inherent randomizing effect of quantum fluctuations. In general, most universes will not turn out warp-free or flat; the distribution of matter will not be nearly smooth; and the pattern of hot and cold spots in the CMB light there will not be nearly scale-invariant. The patches span an infinite number of different possible outcomes, with no kind of patch, including one like our visible universe, being more probable than another. The result is what cosmologists call the multiverse. Because every patch can have any physically conceivable properties, the multiverse does not explain why our universe has the very special conditions that we observe – they are purely accidental features of our particular patch. [Ijjas et al., 2013, pp. 38f.]

According to the 33 authors of a letter [Guth et al., 2017] replying to Ijjas, Steinhardt, and Loeb, the prospects for testing – and ultimately confirming – inflationary cosmology are not nearly as bad as portrayed by the three. We do not have to come to a verdict on that debate in the present context. Clearly, to the degree that specific models of inflation portray it as eternal and leading to a proliferation of island universes, the empirical consequences of inflation must be assessed in the light of this feature.

As explained earlier, due to the proliferation of vacua in the string theory "landscape," the prospects for obtaining any empirical evidence for or against string theory seem independently rather bleak. It would be excellent if that situation could be improved by determining a workable strategy for extracting concrete empirical predictions from multiverse theories such as the landscape multiverse.

1.2.4 Multiverses beyond the Landscape Multiverse

The landscape multiverse scenario is not the only concrete multiverse theory that is seriously considered by at least some physicists. There is a long tradition of cyclic cosmological models that can be seen as proposing multiverse scenarios. These models portray the universe as undergoing many – according to some models, infinitely many – consecutive expansion and contraction cycles, and in some of those models, the parameters characterizing the laws of nature change between cycles. Cyclic cosmologies were popular for a while in the early twentieth century, before Tolman [1934] influentially argued that they are incompatible with basic thermodynamics. Cyclic models have more recently experienced a revival in cyclic brane cosmology, which – like the landscape multiverse – is based on string theory and was developed by Steinhardt and Turok [2001] (see Steinhardt and Turok [2008] for a popular exposition), and in the cyclic model by Baum and Frampton [2007]. Greene [2011, chapters 3–5] provides an accessible popular overview of both the landscape and cyclic multiverse scenarios.

Cosmologist Max Tegmark proposes a categorization of multiverse scenarios into various "levels," which go beyond the level of the cosmological multiverse just discussed. In Tegmark's scheme, the level I multiverse is an infinite extension of the space-time region to which we have causal access. It is not really a multiverse scenario in the sense of the characterization given before because the laws and constants are assumed to be uniform in it.

Tegmark's level II multiverse is a collection of (real) level I multiverses – i.e., different infinite universes – now with different constants and different higher-level physical laws. The landscape multiverse and the cyclic multiverse scenarios just mentioned are level II–type multiverses in Tegmark's sense. Another level II–type multiverse would be one where the Standard Model of elementary particle physics holds in all universes.

Tegmark endorses the Everett interpretation of quantum theory, according to which physical reality is completely described by a universal quantum state. This state undergoes perpetual "branching" into contributions that show almost no mutual inference and effectively function as different "worlds." The totality of the different branches in Everettian quantum theory is Tegmark's level III multiverse.

Tegmark's enthusiasm for multiverse scenarios even extends to an additional, level IV, multiverse, which consists of all *mathematical structures*. Tegmark argues that these are not only mathematically real but also physically real and exist as universes that can be seen as together forming an overarching multiverse, which goes far beyond the level II and III multiverses. As I argue in the penultimate chapter of this book, to the extent that more radical multiverse proposals such as Tegmark's level III and IV multiverse and the philosopher David Lewis's multiverse of *possible worlds* are coherent at all, they cannot be reasonably believed by someone who takes them entirely seriously.

1.3 Overview of This Book

The structure of the remaining chapters of this book is as follows: Chapter 2 reviews the considerations in the physics literature according to which many aspects of the laws, constants, and boundary conditions seem fine-tuned for life as well as criticisms of those considerations. There have been debates about which notion of probability might be best suited to underwrite the claim that fine-tuned parameters are somehow "improbable." Ultimately, the most promising candidate notion of probability to underwrite the claim that life-friendly parameters are "improbable" turns out to be subjective probability, assigned from the perspective of an agent who temporarily abstracts from her knowledge that the parameters have the values that they happen to have.

Chapter 3 discusses a popular alternative response to the finding that many parameters require fine-tuning for life – namely, the inference to some divine designer or God. This response is interesting in itself, and investigating it is an ideal training ground for the discussion of the fine-tuning argument for the multiverse in later chapters. The core idea of the fine-tuning argument for a designer is that a cosmic designer, assuming she/he exists, might for some reason welcome the existence of life and would therefore *set* the parameters as life friendly (something, a further assumption goes, that she is capable of). I discuss how to best formulate this suggested inference from fine-tuning to a designer and consider some objections.

The chief difficulty for the fine-tuning argument for a divine designer, as I argue, is that there are various mutually incompatible versions of the designer hypothesis – some featuring an anthropomorphic designer, others a more abstract one who bears little resemblance to the living organisms we are familiar with. When we consider the latter versions, it seems difficult to see why – and in which sense in the first place – a designer who is so radically unlike any being we are familiar with would favor the existence of life at all and how we could possibly predict his/her "behavior" with some confidence. On a version of the designer hypothesis that features a more anthropomorphic designer, in contrast – here, one might think of a designer along the lines of those encountered in traditional religions – the preferences and actions of the designer may be easy to understand and predict. Unfortunately, such versions of the designer hypothesis are arguably discredited in our post-Darwinian scientific environment and do not deserve serious consideration in the first place. I conclude that inference to a divine designer is probably not an attractive response to the considerations according to which life requires fine-tuning.

Chapter 4 turns to the standard argument from fine-tuning for life to a multiverse as outlined in Section 1.1. This argument, as remarked, centers around the statement that if there is a sufficiently diverse multiverse where the parameters differ between universes, it is no longer surprising that we, as living organisms, exist despite the fine-tuning of the parameters that this requires.

There is a much-discussed objection against the inference from fine-tuning to a multiverse suggested by this argument, namely, that it commits the so-called *inverse gambler's fallacy*: inferring from the existence of one remarkable outcome – in this case, a life-friendly universe – that there are likely many other events – in this case, other universes – most of them with much less remarkable outcomes. To determine whether the inference from fine-tuning to a multiverse really commits this fallacy, I discuss a variety of examples that are structurally similar to the fine-tuned universe and in which it is uncontroversial whether the inverse gambler's fallacy is committed by reasoning that mirrors the fine-tuning argument for the multiverse. In the end, it turns out to be impossible to either affirm or reject the inverse gambler's fallacy charge against the fine-tuning argument for the multiverse. The analogies are either too imperfect or beset by the same ambiguities as the fine-tuned universe. I conclude the chapter by putting forward the suspicion that established standards of rationality may just not determine at all whether the inference from fine-tuning to a multiverse commits the inverse gambler's fallacy or not.

Chapter 5 reformulates the standard fine-tuning argument for the multiverse using the Bayesian language of subjective probabilities. Use of this language allows us to state and address a worry put forward by Juhl [2007]: that belief in some multiverse theory based on the standard fine-tuning argument for the multiverse would inevitably be based on fallacious *double counting* of the fine-tuning evidence. The Bayesian formalism also allows us a fresh look at why it is so difficult to determine whether the standard fine-tuning argument for the multiverse is fallacious or not. As it turns out, that difficulty turns on an ambiguity that relates to the fact that it has been left unclear against which kind of background knowledge the evidence that the parameters are right for life is assessed.

Chapter 6 proposes an alternative, *new* fine-tuning argument for the multiverse, which is by construction structurally immune to the inverse gambler's fallacy charge. The new argument takes a leaf from classic instances of so-called *anthropic* reasoning, as influentially championed by astrophysicists Robert Dicke and Brandon Carter in their accounts of certain *large number coincidences* in cosmology. Since one of Carter's "anthropic principles" is often invoked in expositions of the standard argument from fine-tuning for the multiverse, Dicke/Carter-type anthropic reasoning and the fine-tuning argument for the multiverse are sometimes viewed as instances of the same type of reasoning. But, as it turns out, there is a profound difference between the two in that the fact that we exist (as forms of life) is used as background knowledge in Dicke/Carter-style reasoning, whereas it is used as the evidence whose significance is assessed in the standard fine-tuning argument for the multiverse.

The basic idea of the new fine-tuning argument for the multiverse is that the fine-tuning considerations contribute to a partial erosion of the main theoretical advantage that empirically adequate single-universe theories tend to have over empirically adequate multiverse theories: namely, that their empirical consequences are far more specific. If the parameters require fine-tuning to be compatible with the (long-known) existence of life, this means that all living organisms can only observe very specific values, whether there is only a single universe or a multiverse. It is then no longer a very specific achievement of empirically viable single-universe theories that they correctly predict those values. As a consequence, empirically viable single-universe theories become comparatively less attractive when compared with empirically viable multiverse theories, provided that the fine-tuning considerations do not, along other lines, make belief in a multiverse less attractive.

Chapter 7 finally turns to the prospects for empirically testing specific cosmological multiverse theories such as the landscape multiverse scenario or cyclic multiverse models. The most commonly pursued strategy to extract concrete empirical consequences from specific multiverse theories is to regard them as predicting what "typical" multiverse inhabitants observe if the theories are correct, where "typical" is spelled out as "randomly selected from some suitably chosen reference class." Bostrom [2002] influentially dubbed this principle the *self-sampling assumption*. I scrutinize a proposal by Srednicki and Hartle to treat the self-sampling assumption and the reference class to which it is applied as matters of empirical fact that are themselves amenable to empirical tests. Unfortunately, this proposal turns out to be incoherent.

A much better idea, which coheres well with the intuitive motivation for the self-sampling assumption, is that we should make this assumption with respect to some reference class of observers precisely if our background information is consistent with us being any of those observers and neutral between them. I call this principle the *background information constraint* (BIC) and point out that it at least formally solves the problem of selecting the appropriate observer reference class.

As discussed in Chapter 8, however, applying the BIC in practice is far from straightforward and fraught with difficulties because it requires the regularization of space-time infinities by implementing some cosmic "measure." Furthermore, a suitable physical quantity must be chosen as proxy for the number of reference class observers in some given space-time region. Unfortunately, the choices made by the researchers in this procedure are prone to being exploited – often unintentionally – by the researchers as so-called *researcher degrees of freedom* (this is a term from the social science literature) to yield those results that would best conform to the researchers' theoretical preferences. In the light of this difficulty, the prospects for obtaining compelling evidence in favor of any specific multiverse theory by testing whether our observations are those that typical multiverse inhabitants would make do look bad. As it turns out, the multiverse theories that have the best chances of being successfully tested empirically are those that do not behave as typical multiverse theories in important respects – i.e., those according to which all universes are similar or identical in empirically testable ways.

Chapter 9 starts with the observation that the self-sampling assumption can be seen as an *indifference principle of self-locating belief*: it instructs us to treat all the possibilities who we might be in the reference class of observers as equally likely. Indifference principles of self-locating belief are regarded as suspect by some philosophers, however, because they appear to have paradoxical consequences when applied to certain intensely discussed problems of self-locating belief. Notably, an indifference principle of self-locating belief is usually appealed to in the notorious *Doomsday Argument*, and it also plays a role in the derivation of apparent *anomalous causal powers* in Nick Bostrom's *Adam and Eve* thought experiments. The recommendation that we should sometimes act as if there were anomalous causal powers seems very hard to accept. I show that reasoning akin to that used in the

Doomsday Argument and in the Adam and Eve thought experiments leads to a similar recommendation in a version of the famous *Sleeping Beauty* problem.

All these unattractive recommendations can be avoided if, as required by the BIC, one pays careful attention to the background evidence based on which one assigns probabilities to competing hypotheses and chooses the observer reference class in accordance with that background evidence. I also show that, if one adopts this strategy, one can avoid the unattractive view that the Everett interpretation of quantum theory is confirmed by arbitrary empirical data.

The Everett interpretation is discussed in more detail in Chapter 10, which is also devoted to Tegmark's level IV multiverse of mathematical structures and the multiverse of possible worlds in David Lewis's modal realism.

I discuss the motivation of the Everett interpretation as a possible solution to the foundational problems of quantum theory that allows one to preserve the formalism of the theory without making any amendments or adjustments while being a realist about fundamental physics. The Everett interpretation presents quantum theory as a multiverse theory inasmuch as it postulates many "branches" of reality that supposedly are created by the quantum process of decoherence. It is doubtful, however, whether the appeal to decoherence suffices for the Everettian to identify branches as intended in the absence of a compelling solution to the problem of explaining why quantum probabilities function as self-locating credences in the Everettian multiverse as they do according to the Everettian. I address some suggested solutions to this probability problem – in particular, a recent proposal by Sebens and Carroll [2018] that has received significant attention – and find them wanting.

In the discussion of David Lewis's modal realism, I focus on an old objection against it: that serious belief in it would commit one to inductive skepticism. I argue that, in defending his position against this objection, Lewis unfairly shifts the burden of proof to the proponents of the objection. His thesis entails the existence of epistemic agents for whom inductive inferences will often and radically fail, and the onus is on him to demonstrate that, if his theory is true, we could nevertheless be rationally confident to not be among those agents.

Max Tegmark's thesis that there is a level IV multiverse of mathematical structures, which are all physically realized, hinges on the truth of his *mathematical universe thesis*. According to that thesis, our universe is, itself, a mathematical structure. I argue that this mathematical universe thesis is incoherent because it does not allow us to do justice to important distinctions about which quantities in physical theories do have objective physical significance and which are "surplus formal structure" – e.g., the gauge degrees of freedom in the vector potential formulation of electrodynamics. This undermines Tegmark's case for a level IV multiverse.

The book concludes in Chapter 11 with some reflections on the future of physics in the light of the possibility that we may never be able to, on the one hand, discard multiverse theories as pseudoscience nor, on the other hand, obtain compelling evidence in favor of some multiverse theory or against *any* multiverse theory tout court. What does the future of physics hold if we never find out whether there are other universes with different parameters? I argue that this possibility must be seriously considered: our knowledge of physics "at large" – just as our knowledge of physics "at small" – may forever be limited.

1.4 At a Glance: Some Theses Defended in This Book

The problem of obtaining evidence for or against specific multiverse theories is the central theme of this book. Its central thesis is that there are good reasons for taking multiverse theories seriously, but we may never know if any of them are correct and that this may impede further progress in physics. The more specific stepstone theses argued for while circling around the book's central theme and thesis include the following:

- The fine-tuning argument for a divine designer suffers from the dilemma of having to choose between an anthropomorphic or non-anthropomorphic designer hypothesis. For a non-anthropomorphic designer, the argument is not cogent because the actions of superagents that radically differ from agents with whom we are familiar are impossible to predict; for an anthropomorphic designer, the hypothesis lacks basic scientific credibility.
- Established standards of rationality may not suffice to determine whether the standard fine-tuning argument for a multiverse commits the inverse gambler's fallacy or not.
- A new fine-tuning argument for the multiverse can be formulated, which is structurally similar to the Carter/Dicke anthropic accounts of large number coincidences. That argument is not susceptible to the inverse gambler's fallacy charge, but it requires an independently attractive, empirically viable specific multiverse theory in order to result in rational high degree of belief in a multiverse.
- The observer reference class in cosmology should be chosen in such a way that it includes precisely those observers who we could possibly be, given the background knowledge against which we assess the evidential significance of our observations.
- Derivations of concrete empirical consequences from specific multiverse theories are prone to suffer from confirmation bias. In order to extract such predictions, choices have to be made for which there is no determinately right or wrong option, for instance, concerning the cosmic measure to be used or the physical quantity

used as a proxy for observer number in a given space-time region. This results in so-called researcher degrees of freedom, which researchers are likely to exploit, often unintentionally, in order to obtain results that conform to their theoretical preferences. This problem unavoidably leads to opportunities for confirmation bias to manifest itself, which makes any claimed successful predictions from specific multiverse theories untrustworthy.

- Puzzles of self-locating belief can be resolved by paying attention to the back-ground evidence based on which prior probabilities are assigned and the observer reference class is chosen. Notably, that reference class should be chosen in such a way that it includes every observer that the epistemic agent at issue could possibly be inasmuch as she/he has that background evidence.
- Suggested solutions to the probability problem in Everettian quantum theory are unconvincing. A recent proposal by Sebens and Carroll to derive the Born rule in the Everettian framework using considerations about self-locating belief turns out to suffer from a circularity problem because the central principle on which the derivation is based has no independent motivation beside appeal to the Born rule itself.
- It is not possible to coherently, in full seriousness, believe David Lewis's modal realism. Nor is it possible to coherently believe Max Tegmark's thesis that there is a multiverse of all mathematical structures in which they are physically realized. Tegmark's claim that our universe is itself a *mathematical structure* is found to be incoherent.

All the theses are of independent interest over and above the role that they play, if any, in preparing the ground for the central thesis of the book. Certain aspects of their meaning and their significance become clear only in the context of the arguments that I offer for them in the chapters that follow. So, without further ado, let us jump into those chapters.

2

The Fine-Tuning Considerations

What would our universe have been like if some of its laws, constants, or initial conditions had been somewhat different? According to many physicists, it would probably not have given rise to any interesting complex structures, such as us. Notably, it might not have been hospitable to life in any of its forms.

The fact that our universe is life friendly even though this seems to require a highly special configuration of cosmic parameters is often referred to as its cosmic *fine-tuning*. More generally, the term *fine-tuning* is used to characterize sensitive dependences of facts or properties on the values of certain parameters. Technological devices are paradigmatic examples of fine-tuning. Whether they function as intended depends sensitively on parameters that describe the shape, arrangement, and material properties of their constituents – e.g., the constituents' conductivity, elasticity, and thermal expansion coefficient. Technological devices are the products of actual "fine-tuners" – engineers and manufacturers who designed and built them – but for fine-tuning in the broad sense used in this book to obtain, sensitivity with respect to the values of certain parameters is sufficient.

Various reactions to the universe's alleged fine-tuning for life have been proposed: that it is a lucky coincidence that we have to accept as a primitive given; that it will be avoided by future best theories of fundamental physics; that the universe was created by some divine designer who established life-friendly conditions; and – the reaction that is of special interest for the purposes of this book – that fine-tuning for life indicates the existence of multiple other universes with conditions very different from those in our own universe.

The present chapter reviews the case for various instances of fine-tuning for life (Section 2.1) and it considers whether and, if so, in which sense fine-tuned parameters deserve to be regarded as *improbable* (Section 2.2). Chapter 3 will discuss the suggested inference from fine-tuning for life to a divine designer or God. Taken together, these considerations will be very useful when assessing the suggested inference from fine-tuning for life to a multiverse Chapters 4–6.

2.1 Fine-Tuning for Life: The Evidence

2.1.1 Fine-Tuned Constants

Our best current theories of fundamental physics are the Standard Model of elementary particle physics and the theory of general relativity. The Standard Model accounts for three of the known four fundamental forces of nature – the strong, the weak, and the electromagnetic force – while general relativity accounts for the fourth – gravity. The considerations developed by a number of physicists according to which our universe is fine-tuned for life are aimed at showing that life could not have existed for the vast majority of other forms of the laws of nature, other values of the constants of nature, and other cosmic boundary conditions that characterize our universe in its very early stages. Let us look at some of the claimed instances of fine-tuning to get an impression of the style of reasoning that they exemplify. It should be noted that some of the claims reviewed in what follows are controversial. Some of the criticisms that have been brought forward are mentioned, in Section 2.1.4. For more extensive popular overviews of the fine-tuning considerations see Leslie [1989, chapter 2], Rees [2000], Davies [2006], and Lewis and Barnes [2016]; for more technical ones, see Hogan [2000], Uzan [2011], and Barnes [2012].

First, the strength of gravity, when measured in units of the strength of electro-magnetism, has been claimed to be fine-tuned for life [Rees 2000, chapter 3; Uzan 2011, section 4; Lewis and Barnes 2016, chapter 4]. If gravity had been absent or substantially weaker in relation to electromagnetism, galaxies, stars, and planets would not have formed in the first place. Had it been only slightly weaker (and/or electromagnetism slightly stronger), main sequence stars such as the sun would have been significantly colder and would not explode in supernovae, which are the main source of many heavier elements [Carr and Rees, 1979]. If, in contrast, gravity had been slightly stronger, stars would have formed from smaller amounts of material, which would have meant that, inasmuch as still stable, they would have been much smaller and shorter lived [Adams 2008; Barnes 2012, section 4.7.1].

Second, the strength of the strong nuclear force, when measured in units of the strength of electromagnetism, seems also fine-tuned for life [Rees 2000, chapter 4; Lewis and Barnes 2016, chapter 4]. Had it been stronger by more than about 50 %, almost all hydrogen would have been burned in the very early universe [MacDonald and Mullan, 2009]. Had it been weaker by a similar amount, stellar nucleosynthesis would have been much less efficient, and few, if any, elements beyond hydrogen would have formed. For the production of appreciable amounts of both carbon and oxygen in stars, even much smaller deviations of the strength of the strong force from its actual value might be fatal [Hoyle et al., 1953; Barrow and Tipler, 1986, pp. 252–253; Oberhummer et al., 2000; Barnes, 2012, section 4.7.2].

Third, the difference between the masses m_u and m_d of the two lightest quarks – the up- and down-quark – seems fine-tuned for life [Carr and Rees, 1979; Hogan, 2000, section 4; Hogan, 2007]. Partly, the fine-tuning of these masses obtains with respect to the strength of the weak force [Barr and Khan, 2007]. Changes in this mass difference have the potential to affect the stability of the proton and neutron, which are bound states of these quarks, or lead to a much simpler and less complex universe where bound states of quarks other than the proton and neutron dominate. Notably, in a universe where m_d/m_u is a few times larger than in our universe, neutrons would not be stable within nuclei. In a universe where m_d/m_u is only a few percent smaller than in our universe, protons would combine with electrons into neutrons, which means that hydrogen atoms would not be stable. There are also constraints on the absolute values of the two lightest quark masses, see Adams [2019, figure 5] for an overview.

Fourth, the mass of the electron seems somewhat fine-tuned for life, when compared with the proton and neutron masses, which are more than 1,000 times larger. Notably, according to Tegmark [1998], if the electron mass were more than about $1/81$ of the proton mass, stable ordered structures as needed for chemistry or solid-state physics could not exist. Fluctuations of nuclei would no longer be small compared with the distances between atoms, and this would inhibit the stability of ordered structures. Constraints on the mass of the electron are intertwined with constraints on the strength of the electromagnetic force; see Adams [2019, section 2.3] for an overview.

Fifth, the strength of the weak force seems to be fine-tuned for life [Carr and Rees, 1979]. If it were weaker by a factor of about 10, there would have been much more neutrons in the early universe, leading very quickly to the formation of initially deuterium and tritium and soon helium. Long-lived stars such as the sun, which depend on hydrogen that they can burn to helium, would not exist. Further possible consequences of altering the strength of the weak force for the existence of life are explored by Hall et al. [2014].

Sixth, perhaps most famously, the cosmological constant Λ, which characterizes the energy density ρ_V of the vacuum, has been claimed to exhibit fine-tuning for life. On theoretical grounds, which will be outlined in Section 2.2.2, one would expect it to be larger than its actual value by an immense number of magnitudes. (Depending on the specific assumptions made, the discrepancy corresponds to a factor between 10^{50} and 10^{123}.) However, only values of ρ_V a few order of magnitude larger than the actual value are compatible with the formation of galaxies [Weinberg, 1987; Barnes, 2012, section 4.6; Schellekens, 2013, section 3]. This constraint is relaxed if one considers universes with diferent baryon-to-photon ratios and different values of the number Q (discussed below) which quantifies density fluctuations in the early universe Adams [2019, section 4.2].

The actual value of the cosmological constant is puzzling in that it is extremely small compared to the systematically expected "natural" value. Physicist had long expected it to be zero, which is in several respects a theoretically special value. Prospects for explaining that value would have seemed much better. Accounting for the non-zero, very small positive value of the cosmological constant is widely regarded as one of the central challenges in contemporary fundamental physics.

2.1.2 Fine-Tuned Conditions in the Early Universe

Several aspects of the conditions of the very early universe, as far as presently known, seem fine-tuned for life. Notably, the global cosmic energy density ρ in the very early stages of the universe is extremely close to its so-called *critical* value ρ_c. The critical value ρ_c is defined by the transition from negatively curved universes ($\rho < \rho_c$) to flat (critical density $\rho = \rho_c$) to positively curved ($\rho > \rho_c$) universes. Had ρ not been extremely close to ρ_c in the very early universe, life could not have existed: for slightly larger values, the universe would have recollapsed quickly, and time would not have sufficed for stars to evolve; for slightly smaller values, the universe would have expanded so quickly that stars and galaxies would have failed to condense out [Rees, 2000, chapter 6; Lewis and Barnes, 2016, chapter 5]. In addition to the early universe's striking flatness, its extreme homogeneity and isotropy have also being characterized as instances of cosmic fine-tuning. However, as it is very difficult to specify any supposedly preferred metric on space-times in general relativity, it has been questioned by philosophers whether these fine-tuning claims are well defined at all [Curiel, 2014; McCoy, 2015, 2018].

The relative amplitude Q of density fluctuations in the early universe, known to be roughly $2 \cdot 10^{-5}$, also has been argued to be fine-tuned for life [Tegmark and Rees, 1998; Rees, 2000, chapter 8]. If Q had been smaller by more than about one order of magnitude, the universe would have remained essentially structureless since the pull of gravity would not have sufficed to create astronomic structures like galaxies and stars. If, in contrast, Q had been significantly larger, galaxy-sized structures would have formed early in the history of the universe and soon collapsed into black holes.

Finally, inasmuch as the initial entropy of the universe is a well-defined quantity, it seems to have been exceedingly low. According to Penrose [2004], universes "resembling the one in which we live" [p. 343] populate only one part in $10^{10^{123}}$ of the available phase space volume. The consequences of this instance of fine-tuning are enormous because it is directly related to all the temporal asymmetries that we see around us: the thermodynamic arrow of time, the temporal arrow of radiation, and the asymmetry that we remember the past but not the future.

2.1.3 Fine-Tuned Laws

Not only the constants and initial conditions of the universe, but also the form of the laws of physics itself, have been claimed to be fine-tuned for. Three of the four known fundamental forces – gravity, the strong force, and electromagnetism – play key roles in the organization of complex material systems. A universe in which one of these forces is absent – and the others are present, as in our own universe – would most likely not give rise to life, at least not in any form that resembles life as we know it. The fundamental force whose existence is least clearly needed for life is the weak force [Harnik et al., 2006]. According to the most comprehensive analysis made so far, however, a universe without any weak force would most likely be potentially habitable [Grohs et al., 2018]. Further general features of the actual laws of nature that have been claimed to be necessary for the existence of life are the quantization principle and the Pauli exclusion principle in quantum theory [Collins, 2009, pp. 213f.].

2.1.4 Are the Fine-Tuning Claims Exaggerated?

Many criticisms have been launched against the considerations just reviewed according to which some parameters that characterize our universe are fine-tuned for life.

Among the more dubious of those criticisms is Victor Stenger's claim [Stenger, 2011] that the form of the laws of nature is already completely fixed by the reasonable – very weak – requirement that they be "point-of-view-invariant" in that, as he claims, the laws "will be the same in any universe where no special point of view is present" [Stenger, 2011, p. 91]. Luke Barnes points out several respects in which this criticism fails [Barnes, 2012, section 4.1]. Notably, it confusingly identifies point-of-view invariance with nontrivial symmetry properties that the laws in our universe happen to exhibit. And, as Barnes emphasizes, neither general relativity nor the Standard Model of elementary particle physics is without conceptually viable, though perhaps empirically disfavored, alternatives.

A further similarly dubious criticism, also brought forward by Stenger, is that considerations according to which the conditions in our universe are fine-tuned for life routinely fail to consider the consequences of varying more than one parameter at a time. Barnes [2012, section 4.2] responds to this criticism as well, by giving an overview of various studies such as Barr and Khan [2007] and Tegmark et al. [2006], which explore the complete parameter space of (segments of) the Standard Model and arrives at the conclusion that the life-permitting range in multidimensional parameter space is overall extremely small.

But there are also worries about the fine-tuning considerations reviewed in the previous section that must be taken very seriously. These start with the observation that, by referring to "life," these considerations are meant to apply to life in general, not merely human life. Thus, according to the fine-tuning considerations, a universe with arbitrarily different laws, constants, and boundary conditions would almost certainly not give rise to *any* form of life, however different from ourselves. Evidently, this is a very bold and ambitious claim. Two legitimate worries can be raised about its ambitiousness. The first is that it we have no universally accepted definition of "life," which makes it problematic to advance the sweeping claim about the conditions required for life. The second is that we may seriously underestimate life's propensity to appear under different laws, constants, and boundary conditions because we are biased to assume that all possible kinds of life will resemble life as we know it. A joint response to both worries is that the key message of the fine-tuning considerations is how universes with different laws, constants, and boundary conditions would not merely lack the most characteristic life-supporting structures of our own universe but would have much less structure and complexity overall. This would seem to make them life hostile, irrespective of how exactly one defines "life" [Lewis and Barnes, 2016, pp. 255–274].

A proponent of the second worry may accept this reply but counter it by arguing that we are prone to underestimating the multiple different ways in which complex structures such as galaxies, stars, and planets can, in principle, arise. The route to complexity, they may argue, might well be different in universes with different parameters, but complexity itself – and with it the conditions for some form of life – might be realized in far more circumstances than we can imagine. In support of this claim, they can point to various suggested instances of fine-tuning with respect to which certain originally forwarded bold fine-tuning claims have already been significantly scaled down. Notably, this is true about the claimed fine-tuning of the strong force. Originally, scientists had believed that if this force had been just a few percent stronger, hydrogen would have been absent because proton–proton pairs would have become stable, with devastating consequences for life. But more recent research has shown that significant amounts of hydrogen would most likely exist even if proton–proton pairs ("diprotons") had been stable [MacDonald and Mullan, 2009]. Similarly, as recently argued by Adams and Grohs [2017], if the strong force had been somewhat weaker than we find it to be, so that deuterium would not have been stable, the universe would probably still have been habitable. What would have been different are the processes through which heavier nuclei are synthesized.

Astrophysicist Fred Adams is very critical of the claim that our universe exihibits an extreme fine-tuning for life. According to him, all in all, the range of parameters for which the universe would likely be life supporting is quite considerable. Moreover, as he sees it, the universe could be *more* life friendly, not less, if certain parameters had been different in specific ways. Notably, if the vacuum energy had been smaller, the primordial fluctuations (quantified by Q) had been larger, the baryon-to-photon ratio had been larger, the strong force had been slightly stronger, and gravity slightly weaker, there might well have been more opportunities for life to occur than there are in our own universe [Adams, 2019, section 10.3]. If Adams is correct, our universe seems just garden-variety habitable rather than maximally life supporting. When discussing what would be the rational reaction to our universe's fine-tuning for life in the following chapters, the continuing controversy about the degree of that fine-tuning should be kept in mind.

2.1.5 Avoiding Fine-Tuning for Life through New Physics?

Are the claimed instances of fine-tuning perhaps just artifacts of our current, incomplete, understanding of fundamental physics? One can certainly hope that future developments in physical theorizing will reveal principles or mechanisms which explain the life-friendly conditions in our universe.

There are two different types of scenarios of how future developments in physics could realize this hope: first, physicists may hit upon a so-called *theory of everything* according to which, as envisaged by Albert Einstein, "nature is so constituted that it is possible logically to lay down such strongly determined laws that within these laws only rationally completely determined constants occur (not constants, therefore, whose numerical values could be changed without destroying the theory)" [Einstein, 1949, p. 63]. Einstein's idea is that, ultimately, the laws and constants of physics will turn out to be dictated completely by fundamental general principles. This would make considerations about alternative laws and constants obsolete and thereby undermine any perspective according to which these are fine-tuned for life.

Unfortunately, developments in the last few decades have not been kind to hopes of the sort expressed by Einstein. As discussed in Section 1.2, many physicists continue to regard *string theory* as the most promising candidate "theory of everything" on the table in that it potentially offers a unified account of all known forces of nature, including gravity. But according to our present understanding of string theory, as said in Section 1.2, it has an enormous number of lowest energy states, or *vacua*, which would manifest themselves at the empirical level in terms of radically different effective physical laws and different values of the constants. These would be the laws and constants that we have empirical access to, and so string theory does not even come close to uniquely determining the laws and constants in the manner envisaged by Einstein.

A second type of scenario according to which future developments in physics may eliminate at least some fine-tuning for life would be a *dynamical* account of the generation of life-friendly conditions, in analogy to the Darwinian "dynamical" evolutionary account of biological fine-tuning for survival and reproduction by appeal to natural and sexual selection. Inflationary cosmology, as briefly reviewed in Section 1.2, is perhaps the leading candidate example of such an account. It is widely regarded as providing a dynamical explanation of why the total cosmic energy density Ω in the early universe is extremely close to the so-called critical value Ω_c (see Section 2.1.2) – or, equivalently, why the overall spatial curvature of the universe is close to zero – and of why the universe is both homogeneous and isotropic at accessible scales.

As noted earlier, it is somewhat controversial whether the instances of fine-tuning that inflationary cosmology is supposed to explain are indeed well defined. Moreover, it is controversial whether inflationary cosmology succeeds in its explanatory aspirations even under its own terms. (See Steinhardt and Turok [2008] for harsh criticism by two eminent cosmologists and Earman and Mosterín [1999] for criticism by philosophers.) But even if inflationary cosmology provides successful explanations in these cases, we have little reason to expect that similar accounts will be forthcoming for many another constants, initial conditions, or even laws of nature that seem fine-tuned for life: whereas, notably, the critical energy density Ω_c has independently specifiable dynamical properties that characterize it as a systematically distinguished value of the energy density Ω, the actual values of most other constants and parameters that characterize boundary conditions are not similarly distinguished and do not form any clear systematic pattern [Donoghue, 2007, section 8]. This makes it difficult to imagine that future physical theories will indeed reveal dynamical mechanisms that inevitably lead to these values [Lewis and Barnes, 2016, pp. 181ff.]. Notably, for the cosmological constant Λ, the prospects for devising a dynamical account would seem more promising if its observed value were exactly zero rather than a small positive value close to zero.

2.2 Is Fine-Tuning for Life Improbable?

Let us assume that at least some of the fine-tuning claims regarding the laws, constants, and boundary conditions just reviewed are correct. In that case, is it highly remarkable that there is indeed life despite the required fine-tuning? And should we somehow theoretically *react* to life's existence despite the required fine-tuning, for instance, by inferring – as considered Chapter 3 – that there is likely a divine designer who *made* the parameters life friendly or – as considered in Chapters 4–6 – that there might be other universes? A natural suggestion is that life-friendly parameters require a theoretical response because they are in some sense surprising or *improbable*, given the plentiful different ways that the laws, constants, and boundary conditions might, in principle, have turned out. But in which sense of "probable" could the life-friendly parameters that are manifestly realized plausibly qualify as "improbable"?

2.2.1 Which Type of Probability?

The fine-tuning considerations just reviewed are based on investigations of *physical* theories and their parameter spaces. It may therefore seem natural to expect that the relevant probabilities in the light of which fine-tuning for life is improbable will be *physical* probabilities. On closer inspection, however, this idea is not very plausible: according to the standard view of physical possibility, at least, "physical possibility" just means "accordance with the laws and constants," so counterfactual alternative physical laws and constants are physically impossible by definition [Colyvan et al., 2005, p. 329]. Accordingly, alternative laws and constants trivially have physical probability zero, whereas the actual laws and constants have physical probability one.

There might be ways around this verdict if it turns out that the laws and constants that physicists have uncovered so far are merely *effective* laws and constants in the sense of being fixed by some random process in the early universe that might be governed by more fundamental physical laws. The very idea of such a random process seems open to challenge, however (does it make sense to ascribe any probability to the outcome of an inherently unique and singular event?), but if one were to defend it, one could try to apply the concept of physical probability to those effective laws and constants in which the initial random process could, in principle, have resulted [Juhl, 2006, p. 270]. However, as outlined in the previous section, the fine-tuning considerations do not seem to be based on speculations about any probabilistic process that yields life-friendly parameters with low probability, so they do not seem to implicitly rely on the notion of physical probability along these lines.

Attempts to apply some notion of *logical* probability to fine-tuning for life are beset with difficulties as well. Critics argue that, from a logical point of view, arbitrary real numbers are possible values of the constants [McGrew et al., 2001; Colyvan et al., 2005]. According to them, any probability measure over the real numbers as values of the constants that differs from the uniform measure would be arbitrary and unmotivated. The uniform measure itself, however, assigns zero probability to any finite interval. By this standard, the life-permitting range, if finite, trivially has probability zero, which would mean that life-friendly constants are highly improbable whether or not fine-tuning, in the sense of the considerations reviewed in the previous section, is required for life. This conclusion seems extremely difficult to swallow (though Koperski [2005] argues that it is not as unacceptable for proponents of the view that life-friendly conditions are improbable as it may initially seem).

In view of the difficulties that arise in attempts to apply the physical and logical notions of probability to fine-tuning for life, the most promising approach is to apply an essentially *epistemic* notion of probability to fine-tuning [Monton, 2006; Collins, 2009; Barnes, 2018]. According to this approach, life-friendly conditions are improbable in that we would not have rationally expected them from a certain – to be specified – epistemic point of view.

An obvious problem for this view, however, is that life-friendly conditions are, of course, not literally unexpected to us: as a matter of fact, we have long been aware that the conditions are right for life in our universe, so the epistemic probability of life-friendly conditions appears to be trivially one. As Monton [2006] highlights, to make sense of the idea that life-friendly conditions are improbable in an epistemic sense, we must find a way of strategically abstracting from some of our background knowledge, notably from our knowledge that life exists, and assess the probability of life's existence from that perspective.

Views according to which life-friendly conditions are epistemically improbable face the challenge to provide reasons as to *why* – from an epistemic perspective that ignores that life exists – we should not expect life-friendly conditions. One response to this challenge is to point out that there is no clear systematic pattern in the actual, life-permitting combination of values of the constants [Donoghue, 2007, section 8]. This may be taken to suggest that the actual combination of values does not fare well by the standards of elegance and simplicity. Another response is to appeal to the criterion of *naturalness*. I review and discuss this criterion in Section 2.2.3. There are at least two constants of nature – the cosmological constant and the mass of the Higgs particle – for which this criterion would lead one to expect values that differ radically from the ones that are actually measured: *natural* values of the cosmological constant and the Higgs mass would be many orders of magnitude larger than the ones at which observations hint.

Neither elegance nor simplicity nor naturalness *dictates* any specific probability distribution over the values of the constants, let alone any probability distribution over possible forms of the laws. But, one may argue, any probability distribution over values of the constants that respects the *spirit* of these criteria and that is assigned by abstracting from our knowledge that the universe is indeed life-friendly will inevitably assign very low probability to life-friendly conditions.

2.2.2 Does Improbable Fine-Tuning Unavoidably Call for a Response?

Even if fine-tuned conditions are improbable in some substantive sense, it might be wisest to regard them as primitive coincidences that we have to accept without resorting to such speculative responses as divine design or a multiverse. It is indeed uncontroversial that being improbable does not by itself automatically amount to requiring a theoretical response. For example, any specific sequence of outcomes in a long series of coin tosses has low initial probability (namely, 2^{-N} if the coin is fair, which approaches zero as the number N of tosses increases), but one would not reasonably regard any specific sequence of outcomes as calling for some theoretical response – e.g., a reassessment of our initial probability assignment.

The same attitude is advocated by Gould [1983] and Carlson and Olsson [1998] with respect to fine-tuning for life. Leslie concedes that improbable events do not, in general, call for an explanation, but he argues that the availability of reasonable candidate explanations of fine-tuning for life – namely, the design hypothesis and the multiverse hypothesis – suggests that we should not "dismiss it as how things just happen to be" [Leslie, 1989, 10]. Views similar to Leslie's are defended by van Inwagen [1993], Bostrom [2002, 23–41], and Manson and Thrush [2003, 78–82].

Cory Juhl [2006] argues along independent lines that we should not regard fine-tuning for life as calling for a response. According to Juhl, forms of life are plausibly "causally ramified" in that they "causally depend, for [their] existence, on a large and diverse collection of logically independent facts" [Juhl, 2006, 271]. He argues that one would expect "causally ramified" phenomena to depend sensitively on the values of potentially relevant parameters such as, in the case of life, the values of the constants and boundary conditions. According to him, fine-tuning for life, therefore, does not require "exotic explanations involving super-Beings or super-universes" [Juhl, 2006, 273].

The sense in which fine-tuning for life *fails* to be surprising according to Juhl differs from the sense in which it *is* surprising according to authors such as Leslie, van Inwagen, Bostrom, Manson and Thrush: while the latter hold that life-friendly conditions are rationally unexpected from an epistemic point of view that sets aside our knowledge that life exists, Juhl holds that – *given* our knowledge that life exists and is causally ramified – it is unsurprising that life depends sensitively, for its existence, on the constants and boundary conditions.

2.2.3 *Fine-Tuning and Naturalness*

The notion of *naturalness* has played a central role in model and theory building in fundamental physics in the past few decades. However, as demonstrated in a compelling historical analysis [Borrelli and Castellani, 2019], it has never been a sharp and homogeneous notion. Here I take its underlying motivation to be that the phenomena described by a physical theory should not exhibit an extreme sensitivity with respect to details of a more fundamental (currently unknown) theory to which it is supposed to be an effective low-energy approximation.

Characterizing Naturalness

To understand the motivation behind the naturalness idea as just sketched, it is important to note that modern physics regards our currently best theories of particle physics as collected in the Standard Model as *effective field theories.* Effective field theories are low-energy effective approximations to hypothesized more fundamental physical theories whose details are currently unknown. An effective field theory has an inbuilt limit to its range of applicability, determined by some energy scale Λ^C. Beyond that scale, phenomena become relevant that are covered only by the more fundamental theory to which the effective field theory supposedly is a low-energy approximation. For the theories collected in the Standard Model, it is known that they cannot possibly be empirically adequate beyond energies around the Planck scale $\Lambda_{\text{Planck}} \approx 10^{19}$ GeV, where – presently unknown – quantum gravitational effects become relevant. However, the Standard Model may well be empirically inadequate already at energy scales significantly below the Planck scale. For example, if there is some presently unknown particle with mass M smaller than the Planck scale Λ_{Planck} but beyond the range of current accelerator technology, and if that particle interacts with particles described by the Standard Model, the cutoff scale Λ^C where the Standard Model becomes inadequate may well be M rather than Λ_{Planck}.

In an effective field theory, any physical quantity g_{phys} can be represented as the sum of a so-called bare quantity g_0 and a contribution Δg from vacuum fluctuations corresponding to energies up to the cutoff Λ:

$$g_{\text{phys}} = g_0 + \Delta g. \tag{2.1}$$

The bare quantity g_0 can be regarded as a black box that sums up effects associated with energies beyond the cutoff scale Λ^C where unknown effects must be taken into account. Viewing a theory as an effective field theory means viewing it as a self-contained description of phenomena up to the cutoff scale Λ^C. This perspective suggests that one may only consider an effective theory as *natural* if no delicate cancelation between g_0 and Δg to many orders of magnitude is required in order for the physical quantity g_{phys} to be of its actual order of magnitude. Since the bare quantity g_0 sums up information about physics beyond the cutoff scale Λ^C, if there were such a delicate cancelation between g_0 and Δg, the order of magnitude of the physical quantity g_{phys} would be different if phenomena associated with energies beyond the cutoff scale Λ^C had been slightly different.

One can characterize violations of naturalness as instances of *fine-tuning* in that, where naturalness is violated, low-energy phenomena depend sensitively on the details of some unknown fundamental theory concerning phenomena at very high energies [Williams, 2015]. Physicists have developed ways of quantifying fine-tuning in this sense [Barbieri and Giudice, 1988], critically discussed by Grinbaum [2012].

One way of motivating naturalness as just characterized is by stipulating a probability distribution over values of the bare quantity g_0. Unless that distribution happens to be strongly peaked close to the inferred value of g_0 (which will depend on the cutoff scale Λ^C), the probability for the physical value g_{phys} to be of its actual order of magnitude is tiny. Unnatural values would be *improbable* in that sense. And inasmuch as a physical theory is empirically adequate only for certain unnatural values, this may be seen as speaking against it.

However, whether or not one regards this motivation for naturalness in terms of probability distributions over values of the bare quantities g_0 as compelling may depend on whether one attributes any physical significance to those quantities. It is not clear that one should: for given physical values g_{phys}, the specific values of the bare quantities depend on the chosen regularization scheme and, if a UV cutoff scheme is used, on the scale Λ. Christof Wetterich takes this to mean that the "[f]ine tuning of bare parameters is not really the relevant problem: we do not need to know the exact formal relation between physical and bare parameters (which furthermore depends on the regularization scheme), and it is not important if some particular expansion method needs fine tuning in the bare parameters or not" [Wetterich, 1984, p. 217]. He reiterates this point in more recent work, where he further characterizes the need to fine-tune bare parameters as reflecting "purely a shortcoming of the perturbative expansion series" [Wetterich, 2012, p. 573].

As pointed out by Rosaler and Harlander [2019], Wetterich's criticism of naturalness as a no-fine-tuning criterion turns on an understanding of quantum field theories as defined by entire renormalization trajectories $g^i(\Lambda)$ – i.e., curves in the space of the parameters g^i that are parametrized by Λ. On Wetterich's understanding, hypothetical probability distributions defined over the values $g^i(\Lambda)$ associated with any specific high-energy scale Λ are irrelevant because no such scale is physically preferred. Rosaler and Harlander suggest that Wetterich's understanding might indeed be the appropriate one with respect to high-energy physics. Unlike in condensed matter field theory, which has non-zero minimal inter-particle spacings, there may not be any candidate minimal length scale corresponding to the inverse of some Λ at which the theory would be defined.

Whether or not one finds Wetterich's perspective compelling or not, any motivation for naturalness that relies on an appeal to generic probability distributions is not attractive if one is interested in naturalness primarily with respect to the question of fine-tuning for life: claiming that life-friendly parameters are improbable because they are unnatural, and then accounting for what makes natural values natural in terms of probability would be circular. For the purposes of the present discussion, which focuses on the potential significance of naturalness with respect to the parameters' fine-tuning for life, we must search for another motivation for naturalness.

What is the historical track record of appeals to the naturalness principle? It is sometimes claimed that it had a clear-cut empirical success by enabling the prediction of the charm quark mass. Glashow et al. [1970] had proposed a mechanism that allowed one to explain certain otherwise puzzling observed features of decays governed by the weak nuclear interaction. That mechanism entailed the existence of a new quark, the charm quark, for which there was otherwise no evidence. As Gaillard and Lee [1974] showed, that quark had to have a mass of approximately 1.5 GeV in order to avoid a cancelation between contributions from the strange quark and the charm quark for which there is no independent systematic reasons. Their argument is often reconstructed as an early appeal to naturalness, but this analysis has recently been disputed [Carretero-Sahuquillo, 2019].

All in all, the track record of the naturalness principle in the past few decades is not encouraging. It played a large role in motivating physicists to develop supersymmetric and other extensions of the Standard Model (see the next subsection, "Violations of naturalness: examples," for examples), but no independent empirical evidence for such extensions of the Standard Model has been found, despite intense attempts by experimentalists. In the light of these experiences, eminent particle physicist and long-time advocate of naturalness Gian Giudice speaks of a "dawn of the post-naturalness era" [Giudice, 2017].

There are additional attributions of *hypothetical* successful predictions to the naturalness criterion. Notably, James Wells [2015] claims that a very restrictive criterion of naturalness, which he dubs *absolute naturalness*, would have successfully predicted the existence of the Higgs boson, had it been applied to quantum electrodynamics. A theory is natural in that *absolute* sense if and only if it can be formulated using dimensionless numbers that are all of order one. It is, however, somewhat questionable whether such reconstructed, retrospective successes should be given much weight when assessing the value of naturalness.

Somewhat more permissive than absolute naturalness is 't Hooft's *technical naturalness* criterion ['t Hooft, 1980]. According to that criterion a theory is natural if it can be formulated in terms of numbers that are either of order one or very small but such that, if they were exactly zero, the theory would have an additional symmetry. The motivation for this prima facie arbitrary criterion is that it elegantly reproduces verdicts based on the preceding formulation of naturalness according to which low-energy phenomena should not depend sensitively on the details of some more fundamental theory with respect to high energies. The most prominent violations of naturalness, reviewed in what follows, seem to count as such by the standards of all these different formulations.

Violations of Naturalness: Examples

A prime example of a violation of naturalness occurs in quantum field theories with a spin zero scalar particle such as the Higgs particle. In this case, the dependence of the squared physical mass on the cutoff Λ^C is quadratic:

$$m^2_{H,\,phys} = m^2_{H,0} + \Delta m^2 = m^2_{H,0} + h_t(\Lambda^C)^2 + \dots. \tag{2.2}$$

The physical mass of the Higgs particle is empirically known to be $m_{H,phys} \approx 125$ GeV. The dominant contribution to Δm^2, specified as $h_t(\Lambda^C)^2$ in Eq. (2.2), is due to the interaction between the Higgs particle and the heaviest fermion, the top quark, where h_t is some parameter that measures the strength of that interaction. Given the empirically known properties of the top quark, the factor h_t is of order 10^{-2}. Due to its quadratic dependence on the cutoff scale Λ^C, the term $h_t(\Lambda^C)^2$ is very large if the cutoff scale is large. If the Standard Model is valid up to the Planck scale $\Lambda_{Planck} \approx 10^{19}$ GeV, the squared bare mass $m^2_{H,0}$ and the effect of the vacuum fluctuations would have to cancel each other out to about 34 orders of magnitude in order to result in a physical Higgs mass of 125 GeV. There is no known physical reason why the effects collected in the bare mass term m^2_H should be in such a delicately balance with the effects from the vacuum fluctuations collected in Δm^2. The fact that two fundamental scales – the Planck scale and the Higgs mass – are so widely separated from each other is referred to as the *hierarchy problem*. As a consequence of this problem, the violation of naturalness due to the Higgs mass is so severe.

Various solutions to the naturalness problem for the Higgs mass have been proposed in the form of theoretical alternatives to the Standard Model. In supersymmetry, contributions to Δm^2_H from supersymmetric partner particles can compensate the contribution from heavy fermions such as the top quark and thereby eliminate the fine-tuning problem. However, supersymmetric theories with this feature appear to be disfavored by more recent experimental results, notably from the Large Hadron Collider [Draper et al., 2012].

Other suggested solutions to the naturalness problem for the Higgs particle include so-called *Technicolor* models [Hill and Simmons, 2003], in which the Higgs particle is replaced by additional fermionic particles; models with large extra dimensions, where the hierarchy between the Higgs mass and the Planck scale is drastically diminished [Arkani-Hamed et al., 1998]; and models with so-called *warped* extra dimensions [Randall and Sundrum, 1999].

An even more severe violation of naturalness is created by the cosmological constant ρ_V, which specifies the overall vacuum energy density. Here the contribution due to vacuum fluctuations is proportional to the fourth power of the cutoff scale Λ^C:

$$\rho_V = \rho_0 + c(\Lambda^C)^4 + \dots. \tag{2.3}$$

The physical value ρ_V of the cosmological constant is empirically found to be of order $\rho_V \sim 10^{-3}$ eV. The constant c, which depends on parameters that characterize the top quark and the Higgs particle, is empirically known to be roughly of order one. If we take the cutoff to be of the order of the Planck scale $\Lambda \sim 10^{19}$ GeV, the bare term ρ_0, must cancel the contribution $c(\Lambda^C)^4$ to more than 120 orders of magnitude. Even if we assume a cutoff as low as $\Lambda^C \sim 1$ TeV – i.e., already within reach of current accelerator technology – we find that a cancelation between ρ_0 and $c(\Lambda^C)^4$ to about 50 digits remains necessary. Contrary to the case of the Higgs mass, there are few ideas of how future physical theories might be able to avoid this problem.

Violations of Naturalness and Fine-Tuning for Life

Violations of naturalness are often described as instances of "fine-tuning" in that they involve a sensitive dependence of phenomena associated with large length scales on phenomena associated with very small lengths scales. This type of fine-tuning does not have an immediate connection with the claimed instances of fine-tuning for life discussed in Section 2.1. However, one can construct such a connection and argue that naturalness is relevant to the considerations on fine-tuning for life, along the following lines.

One way of interpreting Eqs. (2.2) and (2.3) is by regarding them as suggesting that the actual physical values of the Higgs mass and the cosmological constant are much smaller than the values that one would expect for them in the framework of the Standard Model. Notably, if the Higgs mass were of order of the cutoff Λ^C – e.g., the Planck scale – and if the cosmological constant were of order $(\Lambda^C)^4$, the bare parameters would not need to be fixed to many digits in order for the physical parameters to have their respective orders of magnitude, which means that the physical values would be natural. Thus, assuming naturalness and the validity of our currently best physical theories up to the Planck scale, one would expect values for the Higgs mass and the cosmological constants of the same order of magnitude as their vacuum contributions – i.e., values much larger than the actual ones.

With respect to the problem of specifying probability distributions over possible values of physical parameters discussed in Section 2.2.1 naturalness may be taken to suggest that all reasonable such distributions have most of their probabilistic weight close to the *natural* values. As explained, for the Higgs mass and the cosmological constant, the natural values are much larger than the observed ones. It is no wonder, therefore, that those who appeal to fine-tuning for life as evidence for a designer or a multiverse put particular emphasis on those instances of fine-tuning for life that are associated with violations of naturalness, notably the cosmological constant [Susskind, 2005, chapter 2; Donoghue, 2007; Collins, 2009, section 2.3.3; Tegmark, 2014, pp. 140f.].

The relevance of naturalness is put into doubt, however, by its failure as a criterion of theory choice in the past few decades to yield any successful predictions. None of the theoretical constructs (partly) motivated by naturalness have been empirically confirmed neither supersymmetry nor technicolor, nor any extra dimensions. This does not rule out any of these classes of theories, but it does harm physicists' trust in naturalness as a principle of theory choice. Sabine Hossenfelder, one of naturalness's staunchest critics, calls it "a mathematically formulated beauty experiment." And she adds that "[i]ts lack of success does not justify its use even as experience-based" [Hossenfelder, 2014, p. 87].

3

Fine-Tuning for Life and Design

3.1 The Inference from Fine-Tuning for Life to Design as a Version of the Teleological Argument

A classic response to the observation that the laws, constants, and boundary conditions are right for life despite the required fine-tuning is to infer that they were likely *made* life friendly by some cosmic designer. Having power over nature in such a profound way and on such a gigantic scale, that designer must, virtually by definition, be some supernatural agent or God, which means that the inference from fine-tuning for life to the existence of a designer can be seen as a version of the teleological argument for the existence of God.

More traditional versions of the teleological argument focused on biological organisms and their parts. An animal's eyes, for instance, consist of matter that is delicately assembled close to its body's surface and connected via nerves to its brain in precisely such a way that detailed information about incoming electro-magnetic waves is transmitted to the brain. This information, in turn, gives the animal crucial hints about its environment, which it may need to spot dangers and predators, become aware of feeding opportunities, seek out potential mates, and much more. The proper functioning of an animal's vision and many other bodily features depends sensitively on its physiological details. These are thus examples of "fine-tuning" in the general sense of the term.

For a long time, appeal to a designer seemed to be the only reasonable response to organismic fine-tuning, but this changed when Darwin and Wallace pointed out natural (and sexual) selection as evolutionary driving forces that can create apparent design by what are actually "blind" mutation and selection processes. The teleolo-gical argument based on organismic design has thereby lost its credibility, despite relatively recent attempts by some researchers to rehabilitate it. These adherents of the so-called *intelligent design* school of thought argue that there are some specific "fine-tuned" features of organisms for which it can reasonably be ruled out that they are the products of any Darwinian evolutionary development based on natural and sexual selection. For example, Michael Behe [1996] claims that the so-called *flagellum*, a bacterial organ that enables motion, is *irreducibly complex*, by which he means that there is no conceivable sequence of small-scale individual evolutionary steps – each of them compatible with selection pressures at the evolutionary times at issues – that would have the organ as we now find it as its outcome. In a similar vein, William Dembski [1998] argues that, in order to explain various apparently fine-tuned features of organisms, evolutionists must hypothesize certain evolutionary steps that are so improbable that one what would not rationally expect them to occur even once in a volume the size of the visible universe. The only alternative explanation that Behe and Dembski regard as available is to infer that these features were produced by an intelligent designer, and so they conclude that such a designer, indeed, likely exists.

The overwhelming consensus in modern biology is that the arguments brought forward by Behe, Dembski, and others do not suffice to revive the organismic design version of the teleological argument. Evolutionary biologist Kenneth Miller [1999] took a detailed look at various examples put forward by Behe as purportedly "irreducibly complex" and found them wanting. For example, as Miller argues, there is, in fact, strong evidence for a Darwinian step-by-step evolutionary history of the flagellum and its constituents [Miller, 1999, pp. 147–148].

With the organismic design version of the teleological argument off the table, the stakes are correspondingly high – at least for the theist – whether the cosmic design version, based on the instances of fine-tuning reviewed earlier, does succeed. By assessing this question, which is interesting in itself, we can learn useful lessons for the discussion of the fine-tuning argument for the multiverse in the two following chapters.

3.2 The Fine-Tuning Argument for Design Using Probabilities

The argument from fine-tuning for life to a divine designer is primarily a suggested inference to the best explanation. If there is a designer – it suggests – she/he *would* create life-friendly parameters even if that requires excessive fine-tuning, so the designer hypothesis supposedly qualifies as a candidate explanation of why there is life despite the required fine-tuning.

To clarify the structure of the argument from fine-tuning for design – and, in particular, to understand the objections that have been levelled against it – it is useful to use the Bayesian language of subjective probabilities to reformulate it, as done by many proponents of the argument [Holder, 2002; Craig, 2003; Swinburne, 2004; Collins, 2009; Manson, 2009]. In general, when performing a Bayesian reconstruction of some argument, one must distinguish between background knowledge and incoming evidence whose evidential impact is assessed. Here we will first look at a Bayesian reconstruction of the fine-tuning argument for design that considers the rational impact of the observation R: that the parameters are right for life. The information that life-friendly parameters are highly special – i.e., that life requires fine-tuning for its existence – is treated as part of the background knowledge.

The hypothesis with respect to which the impact of the evidence R is assessed is the design hypothesis D: that there is a cosmic designer. According to standard Bayesian conditioning, our posterior degree of belief $P^+(D)$ *after* taking into account R is given by our prior conditional degree of belief $P(D|R)$. Analogously, our posterior $P^+(\neg D)$ that there is no cosmic designer is given by our prior conditional degree of belief $P(\neg D|R)$. Bayes's theorem, a trivial identity that follows from the definition of conditional probability, entails that the ratio between the two posteriors is

$$\frac{P^+(D)}{P^+(\neg D)} = \frac{P(D|R)}{P(\neg D|R)} = \frac{P(R|D)}{P(R|\neg D)}\frac{P(D)}{P(\neg D)}. \tag{3.1}$$

Proponents of the argument from fine-tuning for design argue that, in view of the required fine-tuning, life-friendly conditions are highly improbable if there is no divine designer. Thus, the conditional probability $P(R|\neg D)$ should be set close to zero. In contrast, it is highly likely, according to them, that the constants are right for life if there is, indeed, a designer. Thus, the conditional probability $P(R|\neg D)$ should be assigned a value not far from one. *If* a sufficiently powerful divine being exists – the idea goes – it is only to be expected that she/he will be interested in creating, or at least enabling, intelligent life, which means that we can expect the constants to be right for life on that assumption. This motivates the likelihood inequality

$$P(R|D) > P(R|\neg D), \tag{3.2}$$

which expresses the statement that life-friendly conditions confirm the designer hypothesis.

One of the most eminent philosophers to weigh in on the fine-tuning argument for design is Elliott Sober [2003]. For him, only the conditional probabilities as they appear in Eq. (3.2), the so-called likelihoods, can reasonably be assessed, not the unconditional, prior probabilities that appear on the right-hand side of Eq. (3.1). For "likelihoodists" like Sober, the probabilistic version of the fine-tuning argument for design concludes with Eq. (3.2).

Most philosophers who rely on the Bayesian formalism focus not only on likelihoods but also on the priors (and the posteriors resulting from them by considering the likelihoods). For them, the main message of Eq. (3.2) is that, unless the priors $P(D)$ and $P(\neg D)$ dramatically favor $\neg D$ over D in that $P(\neg D) \gg P(D)$, the posteriors will likely favor design in that $P^+(D) > P^+(\neg D)$. Bayesian proponents of the argument from fine-tuning for design conclude that our degree of belief in the existence of some divine designer should be greater than $1/2$ in view of the fact that there is life, given the required fine-tuning.

In what follows, I consider two important objections against the fine-tuning argument for design: the anthropic objection and an objection that I call the "which design hypothesis?" objection. I shall argue that the anthropic objection can be overcome, but only by embarking on considerations that reveal the "which design hypothesis?" objection to be very serious.

3.3 The Anthropic Objection

To understand the anthropic objection, we must first understand what *anthropic reasoning*, more generally, is. Reasoning qualifies as "anthropic" if it is crucially based on the triviality that we, as observers, could not possibly have existed in conditions that are incompatible with the existence of observers. The label "anthropic" derives from the famous *weak anthropic principle* (WAP), put forward under that name (which is somewhat misleading because the principle is not about humans – "anthropoi" – specifically) by Carter [1974]. It is basically a reminder to never let this triviality slip from view:

[W]e must be prepared to take account of the fact that our location in the universe is *necessarily* privileged to the extent of being compatible with our existence as observers. [Carter, 1974, p. 293] (emphasis due to Carter)

The reason why the weak anthropic principle is important even though it highlights a mere triviality is that our methods of empirical observation are unavoidably *biased* toward detecting conditions that are compatible with the existence of observers: we could not possibly have found conditions that are incompatible with our existence. In most empirical research, this potential bias does not play a role when evaluating observations. But in cosmology, it is worthwhile to keep in mind that, even if life-hostile places vastly outnumber life-friendly places in our universe, we should not be surprised to find ourselves in one of the relatively few places that are life friendly and seek an explanation for this finding, simply because – in virtue of being living organisms – we could not possibly have found ourselves in a life-hostile place. Biases that result from the fact that what we observe must be compatible with the existence of observers are referred to as *observation selection effects*.

The weak anthropic principle highlights the potential importance of observation selection effects concerning our (spatiotemporal) location in the universe, whereas what Carter dubs the *strong anthropic principle* (SAP) does the same with respect to observation selection effects that concern our universe and its parameters as a whole:

[T]he Universe (and hence the fundamental parameters on which it depends) must be such as to admit within it the creation of observers within it at some stage. [Carter, 1974, p. 294]

The SAP is sometimes misinterpreted. Most influential has been the misinterpretation by Barrow and Tipler [1986], who understand the "must" in the SAP along teleological lines. But, as Carter [1974] himself highlights (p. 352), see also [Leslie, 1989, pp. 135–145], the SAP is meant to highlight exactly the same type of bias as the WAP and is literally stronger than the WAP only when conjoined with a version of the multiverse hypothesis.

The so-called *anthropic objection* against the argument from fine-tuning for design argues that that argument breaks down once our being biased due to the observation selection effects emphasized by the weak and strong anthropic principles is taken into account. Elliott Sober [2003, 2009] advocates this objection. According to him, we must take into account the observation selection effect OSE that the anthropic principles remind us of not by considering the likelihood inequality (3.2) but the much more problematic

$$P(R|D, OSE) > P(R|\neg D, OSE). \tag{3.3}$$

According to Sober, the design argument only goes through if this inequality is plausible. And whether it is plausible depends of course on how one spells out OSE. Sober himself paraphrases it as "We exist, and if we exist, the constants must be right" Sober [2003, p. 44]. According to this paraphrase, inequality (3.3) is extremely implausible: our existence as living organisms entails that the constants are right for life, which means that the terms on both sides of (3.3) are trivially one and, hence, equal, so (3.3) does not hold on Sober's analysis.

Fortunately for the proponent of the argument from fine-tuning for a designer, Sober's objection proves "too much." As pointed out by critics, Sober's reasoning delivers highly implausible results when transferred to examples where rational inferences are less controversial. Most famous is Leslie's firing squad [Leslie, 1989, pp. 13f.], in which a prisoner expects to be executed by a firing squad but, to his own surprise, finds himself alive after all the marksmen have fired and wonders whether they intended to miss. The firing squad scenario involves an observation selection effect because the prisoner cannot contemplate his post-execution situation unless he somehow survives the execution. His observations, in other words, are "biased" toward finding himself alive. (Juhl [2007] and Kotzen [2012] offer similar examples.) Sober's analysis, applied to the firing squad scenario, suggests that it would not be rational for the prisoner to suspect that the marksmen intended to miss (unless independent evidence suggests so) because that would mean overlooking the observation selection effect that the prisoner is exposed to. But, as seems hard to deny – and as forcefully argued by Leslie [1989]; Weisberg [2005], and Kotzen [2012] – this recommendation is very implausible. It seems quite clear that the prisoner *is* rationally entitled to conclude (or at least seriously consider) that the marksmen intended to miss.

What has gone wrong with Sober's analysis? As pointed out by Weisberg [2005], paraphrasing the observation selection effect OSE with "We exist, and if we exist, the constants must be right" is, on reflection, not very plausible. Instead, according to Weisberg, we should consider the weaker, purely conditional, statement, "If we exist, the constants must be right" [Weisberg, 2005, p. 819] (Weisberg's wording differs) as the appropriate paraphrase of OSE. If we do so, there is no longer any reason to suppose that the inequality (3.3) fails and the argument from fine-tuning for design appears vindicated inasmuch as the anthropic objection is concerned. (See Sober [2009] for a response and Kotzen [2012] for considerations that effectively vindicate Weisberg's point.)

We can gain further insight into what is wrong with Sober's objection and clarify how the fine-tuning argument for a designer can be defended against it by recalling that the best way we have found for making sense of the idea that life-friendly parameters are, in some sense, *improbable* is by appealing to probability in a subjective, epistemic sense. As already noted, there is a prima facie difficulty with this appeal because, as a matter of fact, we have long known that our universe is life friendly, which means that life-friendly conditions are not *literally* unexpected for us – i.e., not literally epistemically improbable. Unfortunately, there is no widely agreed-upon way of how to assess the evidential impact of already known facts in the Bayesian framework – a difficulty that is widely known as Bayesianism's *problem of old evidence* [Glymour, 1980]. In this case, the old evidence whose impact we are trying to assess is the proposition R: that the constants are right for life.

Often, the complication that R is actually old evidence is ignored in Bayesian expositions of the fine-tuning argument for a designer – perhaps understandably, because that problem, as said, has no widely accepted solution. However, as suggested by Monton [2006] and Collins [2009], the so-called *ur-probability* solution to the problem of old evidence has the potential to provide a framework for staging Bayesian versions of fine-tuning arguments, whether for a designer or the multiverse. Barnes [2018], who denies that the old evidence problem is a genuine problem for Bayesianism at all implicitly takes the ur-probability solution for granted.

The basic idea of the ur-probability solution, also known as the counterfactual solution and advocated notably by Colin Howson [1991], is to obtain some rational probability function P^+ that takes the old evidence E properly into account by considering, first, a hypothetical situation where E is not available, assuming, second, that E is acquired, and then, third, performing ordinary Bayesian conditioning with respect to E. In the version of the ur-probability solution used by Monton, one first considers an agent whose hypothetical epistemic situation resembles our own as closely as possible with the sole exception that she lacks the old evidence E whose impact we are trying to assess. The probability function $P(\ )$ that supposedly captures the epistemic agent's credences in that situation is called an *ur-prior*.

The epistemic agent's credences *after* having properly taken into account E correspond to our looked-for posterior P^+. They are obtained from the ur-prior via standard Bayesian conditioning $P^+(H) = P(H|E)$ for any hypothesis H. Crucially, the ur-prior P may differ from our real prior P^-, which expresses our *actual* credences before we systematically assess the impact of E. In the following chapters, it will be useful to express our epistemic agent's credences $P_t(H)$ at arbitrary times t by conditioning of the ur-prior with respect to the information B_0 that she has at those stages

$$P_t(H) = P(H|B_0). \tag{3.4}$$

The ur-probability solution to the problem of old evidence allows us to rescue – with minimal damage – the Bayesian formulation of the fine-tuning argument for a designer despite the fact that R is old evidence. The only adjustment required is that one interprets P as an ur-prior rather than an actual prior. Crucially, this means that $P(D)$ and $P(\neg D)$ must not in any way reflect awareness of the truth of R because R's evidential impact is supposed to be captured by Bayesian conditioning alone. Somewhat oddly, a hypothetical agent with credences $P(D)$ and $P(\neg D)$ would have to be one whose evidence, on the one hand, resembles the total evidence that we have as strongly as possible while, on the other hand, it should not include the fact that the parameters are right for life. Consequently, this hypothetical agent would have to be one who (at least temporarily) is unaware of her own existence.

Admittedly, this is a strange requirement to make, and it dramatically reflects a more general drawback of the ur-probability solution – namely, that it can be very difficult to specify reasonable ur-priors. Monton acknowledges this by conceding that "it is not always clear what values the ur-probabilities should take, especially when one has to make extreme modifications to one's opinion, by, for example, supposing that one does not fully believe that one exists" [Monton, 2006, p. 416].

As we shall see in the next section, this drawback is severe because it makes assessing the relevance of the universe's fine-tuning for life very difficult. But the ur-probability solution to the problem of old evidence at least offers proponents of the fine-tuning argument for a designer a clear-cut rejection of the anthropic objection: the argument in its standard Bayesian form can be saved, including (3.1) and (3.2), with the sole refinement that one must consistently construe all prior probabilities $P(\cdot)$, conditional and unconditional, as "ur-probabilities" – i.e., rational credences of some counterfactual epistemic agent who is unaware that the constants are right for life. At least formally, this lets the proponent of the fine-tuning argument for a designer off the hook with respect to the anthropic objection.

3.4 Which Design Hypothesis Exactly?

The core idea of the fine-tuning argument for a designer, expressed in the likelihood inequality (3.2), is that life-friendly conditions are highly surprising if there is no designer and at least somewhat more to be expected if there is a designer. Some proponents of the fine-tuning argument for divine design argue that this is a rather low bar.

For example, Rota [2016] argues that the argument is very strong even if we grant the skeptics of the fine-tuning argument that we do not know very well how a designer would act:

At the end of the day, how likely is it that an intelligent designer of the universe would design a life-permitting universe rather than a lifeless one? If I had to guess, I'd say over 1/2. But let's concede as much as possible to the person who will say that we know very little about what a universe designer might want. Very well, shall we estimate $P(E|HD\&K)$ at 1 in 10? 1 in 100? How about 1 in a million? The smaller the number, the weaker the fine-tuning argument will be. In order to rely only on a premise that even a sceptic could agree to, let's be generous and assume $P(E|HD\&K) = 1$ in a billion, i.e., $1/10^9$. The reader may be surprised to learn that, even so, the fine-tuning argument will be exceedingly strong. [Rota, 2016, 119f.]

As Rota sees it, the fine-tuning considerations as reviewed in the first section of the previous chapter suggest ascribing a much, much smaller probability to life-friendly parameters on the assumption that there is no designer. So, according to him, the conditional probability of life-friendly parameters if there is a designer will still be extremely high by comparison – even if there is little about the designer that we can say with confidence.

Similar views, collected by Manson [2018], can be found in Hawthorne and Isaacs [2018, p. 151] and are made by Luke Barnes in a fictional dialogue with his coauthor Geraint Lewis in Lewis and Barnes [2016, p. 341].

What is the motivation for this assessment? Many believers conceive of God in a specific way – namely, as a powerful, yet benevolent, actor, who cares immensely about human affairs and is deeply invested in our existence. Surely, at least this kind of God would be disposed to create a life-friendly universe, it seems. And since we want to make at least some room for such a God in our overall designer hypothesis, doesn't this mean that life-friendly parameters, even if by no means guaranteed, are much more probable under the assumption that there is a designer than under the assumption that there is no designer?

Apparently, whether we can expect a potential designer to design depends very much on the specific *version* of the design hypothesis that we happen to consider, where different versions differ according to the kind of designer that they stipulate. What would really be needed for the fine-tuning argument for divine design to be successful is a version of the designer hypothesis that (i) has at least some independent plausibility and that (ii) hypothesizes a designer about whom we could be confident that she/he would indeed design.

One of the most prominent proponents of the fine-tuning argument for a designer, philosopher Richard Swinburne argues, is a promising candidate version of the design hypothesis that fulfills both criteria – namely, the one according to which the designer is "the God of traditional theism." The latter, according to Swinburne, is "a being essentially eternal, omnipotent (in the sense that He can do anything logically possible), omniscient, perfectly free, and perfectly good" [Swinburne, 2003, p. 107]. (See Swinburne [2004, chapter 8] for Swinburne's most complete exposition of his version of the fine-tuning argument.) He argues that we can be at least moderately confident that the God of traditional theism, if he exists, "will bring about an orderly, spatially extended, world in which humans have a location" [Swinburne, 2003, p. 113]. (Swinburne operates with a generalized, non-biological concept of "humans," according to which intelligent beings that only vaguely resemble us qualify as "humans.") Notably, life-friendly conditions, conditional on the existence of the God of traditional theism, do not have very low probability according to Swinburne; i.e., $P(R|D)$ is not many orders of magnitude smaller than one. Thus, for Swinburne, if we consider traditional theism as our preferred design hypothesis, it seems plausible that the inequality (3.2) obtains.

Swinburne's "God of traditional theism" is eternal, omnipotent, omniscient, perfectly free, and perfectly good, and thereby rather unlike any creature or agent that we are familiar with in our everyday lives. In his sublime perfection, He is also unlike the gods that we encounter in ancient mythology and, for that matter, unlike the god of, say, the Old Testament, who often becomes angry and regrets his own actions – traits that are difficult to reconcile with the ones ascribed to the God of traditional theism by Swinburne. What follows from Swinburne's characterization of his God is that he is very much *unlike* any human agent. Actually, it seems doubtful that this God is coherently conceived of at all – apparently even to Swinburne, who devotes an entire book [Swinburne, 2016] to that question. In any case, in all his superhuman glory, he differs completely from agents with whom we are familiar from this world and is thus very much a creature of what Van Schaik and Michel [2016] call *intellectual* religion, contrasting it with what they call *intuitive* religion.

A serious problem with versions of the designer hypothesis along the lines of intellectual religion, which conceive of the designer in a consistently non-anthropomorphic way, was pointed out long ago by John Venn [1866] and later by John Maynard Keynes [1921]: it is extremely difficult to competently judge what a divine designer who is dramatically unlike human beings would do, if she/he existed. Elliott Sober puts this concisely:

Our judgements about what counts as a sign of intelligent design must be based on empirical information about what designers often do and what they rarely do. As of now, these judgements are based on our knowledge of *human* intelligence. The more our hypotheses of intelligent designers depart from the human case, the more in the dark we are as to what the ground rules are for inferring intelligent design. [Sober, 2003, p. 38]

The same point is made by Jan Narveson [2003], according to whom it is impossible to know how an extremely powerful, highly intelligent being that is very much unlike ourselves would act. As he puts it, "[b]odiless minded super-creators are a category that is way, way out of control" [Narveson, 2003, p. 99]. Proponents of intellectual religion implicitly acknowledge this point when they react to the problem of Evil – the existence of suffering despite God's supposed omnipotence and benevolence – by pointing to, as Narveson puts it, "the mysterious ways of the Deity" [Narveson, 2003, p. 99]. Thus, in order to be convincing, the fine-tuning argument for a designer should not be combined with a highly abstract and intellectualized conception of that designer: it will otherwise be impossible to motivate how the designer *would* act if she/he existed. Put differently, it will otherwise be impossible to rationally motivate assigning a value to the conditional probability $P(R|D)$ that is not extremely close to zero.

In a recent criticism of Rota [2016] and Hawthorne and Isaacs [2018], Manson [2018] agrees with this pessimistic verdict on assessing reasonable values for $P(R|D)$: according to Manson, the intentions of the hypothetical designer are simply *inscrutable* to us: we can only speculate about her/his preferences and intentions. We have simply no basis for confidently assigning *any* numerical value to the probability $P(R|D)$, and so we cannot assume, as the proponent of the fine-tuning argument for design does, that $P(R|D)$ is significantly larger than $P(R|\neg D)$.

I think that the opponent of the design argument can make an even stronger point here than the one made by Manson: not only that it is impossible to for us to confidently assign any specific numerical value to the probability $P(R|D)$, but that we have no reason for systematically differing in our assignments of values to $P(R|D)$. For a "bodiless minded super-creator," as Narveson puts it, whose mind is radically unlike any mind that we are familiar with, there is no reason to assume that she/he would favor the existence of humans over their nonexistence, nor any reason to assume that she/he would care about the existence of humans at all. It is not even clear that the very notion of "caring" or "being interested in applies here."

The same considerations apply to other versions of intellectual religion beyond Swinburne's traditional theism. The anthology [Buckareff and Nagasawa, 2016] features a plethora of "alternative concepts of God," all articulated and defended by professional philosophers of religion, notably versions of pantheism and panentheism. According to some these concepts, inasmuch as they are articulated intelligibly, God does not really have any agent-characteristic traits at all, so whether they provide a candidate designer in the sense of the fine-tuning argument at all is questionable.

There are other, more intuitive and anthropomorphic versions of the designer hypothesis D for which the problem that were are unable to determine what a designer would do if she/he existed does not arise: deities like the goddesses and gods of ancient Greece and Rome, the God of the Hebrew Bible, and the deity/deities of Christianity have many characteristically human traits, and this enables us to understand their actions in terms of human categories and to predict, at least to a certain degree, how they *would* act in certain hypothesized circumstances (e.g., when pondering whether to create beings like us).

The problem with these versions of the designer hypothesis is that they are in conflict with well-established central aspects of the modern scientific world view. "Bodiless minded super-creators," as Narveson calls them, might be sufficiently abstract to be immune against being ruled out by our everyday and scientific knowledge. But we have compelling reasons for believing that human-type superagents like the ancient gods – spatiotemporally positioned, suspiciously similar to us humans, passionate about human affairs, but with certain superpowers that make them less constrained than we ordinary humans are in our actions – do not exist. Any ur-prior $P(D)$ that we can reasonably motivate for such intuitive versions D of the design hypothesis will therefore plausibly be negligible.

A desperate measure to preserve the fine-tuning argument for a designer despite these difficulties is to focus on the highly specific "special design" hypothesis SD, which is tailored to fulfill the likelihood inequality (3.2) by *defining* the designer as a being with both the intention and ability to create a universe exactly like the one that we find ourselves in. But, of course, this version of the designer hypothesis is far too ad hoc and "fitted to the facts" to have sufficient independent motivation and plausibility for deserving serious consideration in the first place. To use Bayesian terms, there is no reason for ascribing it an ur-prior $P(D)$ that differs even minimally from zero.

In fact, having chosen the ur-probability solution to the problem of old evidence in response to the anthropic objection, motivating a non-negligible prior $P(D)$ for *any* version of the designer hypothesis is extremely difficult: for, remember, in the framework of the ur-probability solution, our assumed background knowledge must be restricted to facts that do not entail the existence of life. Collins argues that if we focus only on a limited class of constants C, the background evidence that we can use to motivate the prior $P(D)$ is allowed to "includ[e] the initial conditions of the universe, the laws of physics, and the values of all the other constants except C" [Collins, 2009, p. 243]. But appeals to the sacred texts of religions cannot be used to motivate the ascription of a non-negligible ur-prior $P(D)$ because they presuppose, and thus entail, the existence of life. Notably, as pointed out by Monton, "[i]n formulating an urprobability for the existence of God, one cannot take into account Biblical accounts about Jesus" [Monton, 2006, p. 418].

According to Monton [2006, p. 419], this still leaves the door open for proponents of the argument from fine-tuning for design to motivate a non-negligible ur-prior $P(D)$ by resorting to arguments for the existence of God that are either a priori – e.g., the ontological argument – or appeal only to very general empirical facts that do not entail that the conditions are right for life – e.g., the cosmological argument. But, one may object, those arguments belong to the realm of intellectual religion. The designer whose existence is supposedly motivated by appealing to them will inevitably be totally unlike any concrete human agent. As we have seen, this makes it very difficult for the design theorist to give a compelling reason for assigning a comparatively large conditional probability $P(R|D)$. It is also unclear whether such a designer, if we take seriously how *unhuman* and, in that sense, strange she/he really is, can do justice to the spiritual and emotional needs of a pious religious believer.

3.5 An Alternative Fine-Tuning Argument for Design

The argument from fine-tuning for a designer in the form discussed so far has an awkward feature – namely, that it treats the fact that life requires fine-tuned conditions as background knowledge and assesses the evidential significance of the observation that life-friendly conditions obtain against that background. An alternative argument from fine-tuning for design, explored by John Roberts [2012] and independently investigated and endorsed by White [2011] in a reply to Weisberg [2010], treats our knowledge that the conditions are right for life as background information and assesses the rational impact of physicists' insight that life requires fine-tuned conditions against this background.[1] An advantage of this alternative is that it fits better with our actual epistemic situation: that the conditions are right for life is something we have known for a long time; our actual new evidence is that the laws of physics – as White [2011] and Weisberg [2012] put it – are *stringent* rather than *lax* in the constraints that they impose on the constants and boundary conditions if there is to be life.

---

[1] This version of the argument is closer in spirit to an alternative solution to the problem of old evidence: namely, the one due to Garber [1983], in which Bayesian conditioning is applied not to the old evidence itself but rather to the insight that the theory or hypothesis in question – in our case, the designer hypothesis – supposedly *entails* or, in a modified scheme that was recently proposed by Hartmann and Fitelson [2015], *explains* the old evidence – in our case, the parameters are life friendly. There might be other ways of applying Garber-style solutions to the fine-tuning argument (both the one for a designer and the one for a multiverse) that have not been considered in the literature so far. Notably, one may try to base those attempts on approaches that try to combine the virtues of extant solutions to the old evidence problem – e.g., the approach due to Sprenger [2015].

The central likelihood inequality around which White's version of the argument revolves is

$$P(S|D, O) > P(S|\neg D, O), \tag{3.5}$$

where D is, again, the designer hypothesis, S is the proposition that the laws are stringent (i.e., that life requires delicate fine-tuning of the constants), and O is our background knowledge that life exists [White, 2011, p. 678]. (There is an assumption in Roberts's argument [Roberts, 2012, p. 296] that plays an analogous role as (3.5).) The inequality (3.5) expresses the statement that stringent laws confirm the designer hypothesis, given our background knowledge that life exists. Does it plausibly hold for reasonable probability assignments? White argues that it does and supports this claim by giving a rigorous derivation of (3.5) from assumptions that he regards as plausible. Crucial among them is the inequality

$$P(D|S) \geq P(D|\neg S), \tag{3.6}$$

which White motivates by arguing that "the fact that the laws put stringent conditions on life does not by itself provide any evidence *against* design" [White, 2011, p. 678]. Put differently, according to White, absent information that life exists, information that the laws are stringent does at least not speak *against* the existence of a designer.

But White's argument for (3.6) is unconvincing after all, as pointed out by Weisberg [2012] – who takes his criticism to undermine (3.5) – who shows that it is implausible by the design theorist's own standards. The design theorist holds a combination of views according to which, on the one hand, life is more probable if there is a designer than if there is no designer and life is less probable if the laws are stringent rather than lax. If one adds to this combination of views the assumption that none of the possible life-friendly conditions has higher probability than the others, both if there is a designer and if there is no designer, it dictates that – bracketing knowledge that life exists – stringent laws speak *against* the existence of a designer; i.e., it dictates $P(D|S) < P(D|\neg S)$, contrary to (3.6). Absent any evidence that life exists, evidence that the laws are stringent speaks against the existence of life in that stringent laws make life unexpected.

The design theorist, as anticipated by Weisberg [2012, p. 713]), may respond to this objection and defend (3.6) by arguing that the designer would plausibly *first* choose either stringent or lax laws, sidestepping her intention to enable the existence of life at that stage or actively preferring stringent laws, and only *then* choose life-friendly constants. But this strategy runs afoul of the central difficulty for any version of the designer hypothesis that hypothesizes a sufficiently abstract non-anthropomorphic designer, pointed out in the previous section: that we have little experience with such super-beings and, therefore, are not well positioned to predict their preferences and likely actions *if*, indeed, they exist.

Much more could be said in defense of the fine-tuning argument for a divine designer as well as against it. But the present book, even though it ventures into the territory of philosophy of religion, focuses on the epistemology of fundamental physics in general and multiverse theories in particular. With respect to that focus, we can extract two crucial morals from the present discussion of the fine-tuning argument for divine design, which will both be useful for the discussion of the fine-tuning argument for the multiverse in the two subsequent chapters: first, anthropic considerations are important and relevant in potentially non-obvious ways; and second, turning from some general response to fine-tuning for life – in this case, the designer response – to specific *versions* of that response has the potential to considerably alter the picture. We will discuss in the next two chapters to what degree one can make a case for inferring from fine-tuning for life that *some* version of the multiverse hypothesis is likely correct. Chapters 7 and 8, in turn, will discuss how large the challenge of empirically assessing concrete multiverse theories really is.

Part II

Fine-Tuning for Life and the Multiverse

4

The Standard Fine-Tuning Argument for the Multiverse

Having reviewed the considerations according to which various aspects of the laws, constants, and boundary conditions in our universe appear fine-tuned for life and the suggested inference from these considerations to the existence of a divine designer in the previous chapter, let us now turn to their possible relevance to multiverse theories. For the purposes of this discussion, it is useful to treat these multiverse theories collectively under the banner of a single general "multiverse hypothesis." The defining feature of this hypothesis is that, according to its constituent theories, certain parameters that require fine-tuning for life scan over a broad range of values across different "subuniverses."

4.1 The Inverse Gambler's Fallacy Charge

The standard fine-tuning argument for the multiverse, as briefly introduced in Chapter 1, starts with the idea that, if there is a sufficiently diverse multiverse in which the parameters differ between universes, it is only to be expected that there is at least one where they are right for life. As discussed in the previous chapter, the strong anthropic principle points out that the universe in which we, as observers, find ourselves must be one where the conditions are compatible with the existence of observers. Let us, for the sake of simplicity, assume that observers are often forms of life. This suggests that, if there is a sufficiently diverse multiverse, it is neither surprising that there is at least one universe that is hospitable to life nor that we find ourselves in a life-friendly one. We could simply not have found ourselves in a life-hostile universe. The truth of the general multiverse hypothesis would therefore make it unsurprising why we exist despite the fine-tuning required for the existence of life. Many physicists – e.g., Susskind [2005]; Greene [2011] and Tegmark [2014] – and philosophers – e.g., Leslie [1989]; Smart [1989]; Parfit [1998] and Bradley [2009] – endorse this line of thought and interpret it as providing a strong motivation for responding to the fine-tuning considerations by assuming that there is likely a multiverse.

The most-discussed objection against this argument is that it commits the *inverse gambler's fallacy*, originally identified by Ian Hacking [1987]. This fallacy consists in inferring from an event with a remarkable outcome that there have likely been many more events of the same type in the past, most with less remarkable outcomes. For example, the inverse gambler's fallacy is committed by someone who enters a casino and, upon witnessing a remarkable outcome at the nearest table – say, a fivefold six in a toss of five dice – concludes that the toss is most likely part of a large sequence of tosses. If the outcomes of the tosses can be assumed to be probabilistically independent – which is a reasonable assumption as long as the outcomes neither influence each other nor have a common cause – then this inference is indeed fallacious.

According to critics of the standard argument from fine-tuning for the multiverse, the argument commits this fallacy by, as White puts it, "supposing that the existence of many other universes makes it more likely that *this* one – the only one that we have observed – will be life-permitting" [White, 2000, p. 263]. Versions of this criticism are endorsed by Draper and Pust [2007] and Landsman [2016]. Somewhat curiously, Hacking [1987] himself regards only those versions of the argument from fine-tuning for the multiverse as guilty of the inverse gambler's fallacy that infer the existence of multiple universes in a temporal sequence.

Adherents to the inverse gambler's fallacy charge against the argument from fine-tuning for the multiverse claim that it is misleading to focus on the impact of the proposition R – that the conditions are right for life in *some* universe. As they highlight, much closer to our full evidence is the more specific proposition H: that the conditions are right for life *here*, in *this* universe. According to proponents of the inverse gambler's fallacy charge, if we replace R with H, the fine-tuning argument for the multiverse no longer goes through because finding *this* universe to be life friendly does not increase the likelihood of there being any other universes.

But not all philosophers agree with the inverse gambler's fallacy charge against the fine-tuning argument for the multiverse. Various lines of defense of the inference to a multiverse are developed by McGrath [1988]; Leslie [1988]; Bostrom [2002]; Manson and Thrush [2003] and Juhl [2005]. According to McGrath, the proper analogy is not the one considered by Hacking but the one with an observer who is allowed to enter a casino *only if and when* some specific sequence of outcomes occurs. In that setting, when the observer is allowed to come in, it is indeed rational for her to assume that there have been multiple trials rather than just one, in analogy to the suggested inference from fine-tuning for life to a multiverse. White [2000], in turn, criticizes McGrath's analogy as inadequate because, as he puts it, "it is not as though we were disembodied spirits, waiting for some big bang to produce a universe that could accommodate us" [White, 2000, p. 268].

Darren Bradley [2009], in turn, attacks White's endorsement of the inverse gambler's fallacy charge. According to him, it does not do justice to the *observation selection effect*, which consists in the fact that – since we could not have existed in a life-hostile universe – our observations are biased toward finding constants that are right for life. As Bradley sees it, if we take this effect into account, it becomes clear that the fine-tuning argument for the multiverse does not commit the inverse gambler's fallacy after all. However, Klaas Landsman [2016] has recently disputed the adequacy of Bradley's analogy, and consensus on whether the fine-tuning argument for the multiverse commits the inverse gambler's fallacy or not does not seem within reach. In the following section, I will try to make both perspectives in this debate as transparent as possible by using urn analogies.

4.2 Urn Analogies

Bradley illustrates the fine-tuning argument for the multiverse while defending it against the inverse gambler's fallacy charge in terms of an urn that contains either one or two balls, depending on the outcome of a fair coin toss. Balls come in two different sizes: large (L) and small (S). For each ball, an additional fair coin toss determines whether it is large or small; i.e., $P(L) = P(S) = 1/2$ for each ball. A small hole is opened in the urn through which a small ball is sampled if there is one, whereas a large ball would not fit through the hole. This makes the sampling procedure *biased* toward small balls and introduces an observation selection effect that must be accounted for in one's degrees of belief about how many balls there are in the urn. Indeed, if a small ball is sampled, this information is not neutral with respect to "one ball in the urn" versus "two balls in the urn," which it would be if the sampling procedure were not biased toward small balls. Sampling a small ball confirms "two balls in the urn" over "one ball in the urn" because, given "two balls," it was more likely that the urn would contain at least one small ball (which could be sampled) than given "one ball."

Bradley constructs an analogy between this example and the problem of the apparently fine-tuned constants by treating small balls as symbolizing life-friendly universes and large balls as symbolizing life-hostile ones.[1] According to this analogy, just as we could not possibly sample a large ball from the urn, we could not have possibly found ourselves in a universe that is not life friendly. And just as sampling a small ball from the urn confirms "two balls" over "one ball,"

---

[1] Inasmuch as one regards the probability for a universe to be life friendly as very small, one may want to adjust the example so that the probability for a ball to be small is also very small: $P(S) \ll 1$.

Bradley claims, if we find ourselves in a universe where the constants are right for life, this supports "two universes" over "one universe." If "N balls" (with large $N \gg 1$) is also a possibility, sampling a small ball confirms this even more strongly than "two balls." Analogously, "N universes," corresponding to the multiverse hypothesis, is confirmed even more strongly than "two universes" when we find out that we exist despite the required fine-tuning.

Rational credences in this urn example are uncontroversial, but is the example itself really analogous to the problem of the fine-tuned constants? Klaas Landsman, one of those who believe that the fine-tuning argument for the multiverse commits the inverse gambler's fallacy, claims not and suggests that another urn example provides a much better analogy [Landsman, 2016, section 5]. In Landsman's example, the competing hypotheses are not about how many balls there are in a single urn but about how many *urns* there are, where each urn contains only one ball, large or small, with $P(L) = P(S) = 1/2$ for each urn independently. As before, only small balls can be sampled. Importantly, an observer's observations are confined to a single fixed urn. If she samples a small ball from that urn and concludes that there are likely more urns, she indeed commits the inverse gambler's fallacy. The selection bias in favor of small balls is irrelevant to the rational credences about the number of urns. No observation selection effect must be taken into account.

Landsman argues that this example can be recognized as relevantly analogous to the problem of the fine-tuned constants if we let urns, not balls, stand for universes. For a given urn, the size of its ball indicates whether the universe symbolized by the urn is life friendly or not. According to Landsman, just as we have access to only a single fixed urn in the example, our observations are confined to a single fixed universe. If an observer samples a small ball from her urn, this corresponds to our finding that the constants in our universe are right for life. And just as sampling a small ball from one's urn is uninformative about whether there are more urns, finding the constants right for life in our universe is uninformative about whether there are more universes.

Which analogy is correct, Bradley's or Landsman's? It would be possible to answer this question with confidence only if it were clear whether it is rational for us to reason as if we possibly *could have lived* in a different universe or not. Bradley's analogy assumes that it is rational to reason in that way: in his example, we can sample *any* small ball from the urn (if there is one), so the analogy presupposes that our existence is not tied to a specific universe (corresponding to some specific ball), which may or may not be life friendly. Landsman's analogy, in contrast, can only be adequate if it is *not* rational for us to reason as if we could have lived in a different universe: in his example, our attention is confined to a single urn with a single ball in it (large or small), which means that the analogy presupposes that our existence is tied to a specific universe (corresponding to a specific urn) that may or may not be life friendly.

4.3 Broadening the Debate

So, how should we reason? As one may have guessed (or feared), philosophers diametrically disagree on the answer to this question: White [2000, p. 269] argues that it would be irrational to reason as if we could have existed in a different universe, whereas Manson and Thrush [2003, p. 76f.], Juhl [2005, 345ff.], and Bradley [2009, p. 68] claim that there is no good reason for refraining from doing so.

Complementing his case against White, Bradley [2009, pp. 68–70] presents an argument according to which even if, as White suggests, we should reason as if we could not have existed in a different universe, the existence of many universes would still make it more likely that the particular universe in which we could have existed – our own – indeed exists. White could object, however, that we should account for the life-friendliness of this universe, not its existence – and that its life-friendliness is not made any more likely by the existence of multiple other universes.

Epstein [2017] has recently proposed a new argument for Bradley's position and against Landsman's by arguing that a familiar point about good scientific practice supports it:

The problem here stems from a failure to respect what is sometimes called the "Predesignation Requirement," a widely-accepted rule of sound scientific practice .... It is a familiar point that experimental design should ideally have the following structure. First, a hypothesis is generated. Then, an experimental procedure is designed to test the hypothesis against an alternate hypothesis (often the null hypothesis). The possible outcomes of the procedure are identified, and the evidential import of each possible outcome is assessed: the experimenters determine how their credences in the competing hypotheses ought to be updated, if a given possible outcome is observed. Then the experiment is run, the observation is made, and credences are updated accordingly.

A standard worry in experimental science arises when, instead of following this procedure, experimenters first observe the outcome of the experiment, and then use that outcome to generate hypotheses post hoc. Notoriously, such violations of the Predesignation Requirement – sometimes referred to as "data-mining" or "over-fitting" – can lead to the appearance of statistically significant experimental outcomes when no real confirmation has occurred. [Epstein, 2017, p. 648]

In the case of the inverse gambler's fallacy objection against the fine-tuning argument for the multiverse, Epstein sees this mistake as committed by those who treat the observation that *this* universe is life permitting as the relevant one: this universe was not in any way predesignated before we determined its life-friendliness, but its being life friendly was a precondition for its coming to our attention.

However, as Epstein implicitly acknowledges in this passage where he defends his view, the "process" by means of which we came to focus on our own universe as the object of our observations seems quite peculiar and difficult to capture:

The problem with th[e] suggestion [that we should reason as if we could not have existed in a different universe] is that it in effect treats our observation as an outcome of a process we were observing, where we were focused, in particular, on whether we would come to exist in α [our universe, rigidly designated]. It suggests that the relevant partition of possible outcomes divides those outcomes according to whether or not, as observers of the process in question, we observe α – our predesignated potential "home universe" – to be life-sustaining. But such a picture is no more plausible than the one White mocks: just as "it is not as though we were disembodied spirits, waiting for some big bang to produce a universe that could accommodate us," it also is not as though we were disembodied spirits, keenly observing α – our designated potential home – and hoping that it, in particular, would be able to accommodate us. When considering the probabilistic process that led to our existence as observers, we cannot accurately capture the nature of our evidence by modeling that process as one that we – as the particular beings we are, tied specifically to α – observed. We don't have the proper perspective on such a process to assess its outcomes in terms of the observations we, specifically, could make. [Epstein, 2017, p. 653]

Epstein is right that there was no "process" by means of which we selected α, our universe, to be the object of our study. We did not somehow select our universe prior to knowing whether it contains any life and then found out that it does; but neither did we start out with a collection of universes, not knowing how large that collection is, and then receive the piece of evidence that the collection contains at least one universe that is life friendly.

Thus, the "process" in which we "selected" our universe as the object of our study, in order to subsequently "find out" that it is life friendly is quite unlike more standard sample-selecting procedures in the sciences. Epstein is right in that this makes the inference from fine-tuning for life to a multiverse very different from paradigmatic instances of reasoning that commit the inverse gambler's fallacy. But it arguably also makes it rather different from situations where this fallacy is clearly *not* committed.

Perhaps there are further strong reasons – overlooked or not taken seriously enough by me – to believe that either Bradley's or Landsman's analogy is adequate whereas the other one is inadequate. In the absence of such reasons, the best we may be able to do is to look for further analogies to the inference from fine-tuning for life to a multiverse: analogies that are much more overtly analogous to the problem of the fine-tuned constants than the urn examples.

The following three sections consider three fine-tuning problems that are much more patently analogous to the problem of the fine-tuned constants than the urn and casino scenarios that have been predominantly considered so far in the literature and in this chapter: the problem of our fine-tuned planet, the problem of our fine-tuned ancestors, and the problem of our lucky civilization, introduced and discussed in Sections 4.4–4.6 respectively.

At least in the first of these three analogies, reasoning that parallels the inference from fine-tuning to a multiverse seems prima facie adequate. But the only reason why the inverse gambler's fallacy charge is usually not brought up in it may well be that we have strong independent reasons to believe in the hypothesis that is analogous to the multiverse hypothesis here. In the absence of such independent reasons, it would be coherent, though perhaps not compelling, to raise the inverse gambler's fallacy charge against the inference that parallels the inference from fine-tuning to a multiverse.

Interestingly, in the third analogy, we have no strong independent reasons to believe in the hypothesis that plays an analogous role to the multiverse hypothesis. Tellingly, whether the inference analogous to the inference from fine-tuning to the multiverse is adequate seems much more questionable in this problem.

The ultimate conclusion to this chapter will be the suggestion that we should seriously consider the possibility that we will never be able to either determinately reject or endorse the inverse gambler's fallacy charge against the fine-tuning argument for the multiverse: established standards of rationality may just not allow us to decide whether the inference from fine-tuning to a multiverse commits it or not.

4.4 The Fine-Tuned Planet

The first analogy – the problem of the *fine-tuned planet* – starts with the observation that life could not have appeared and evolved on Earth if Earth's size and mass, its distance from the sun, the size and distance of its neighboring planets, and the abundance of certain chemical elements on it had been significantly different. The parameters that describe these conditions on Earth appear fine-tuned for life in a similar way to how the constants of our universe appear fine-tuned for life. Earth's fine-tuning may seem less dramatic and impressive than the fine-tuning of the constants,[2] but there does not seem to be any principled difference between the two: inasmuch as the life-friendliness of the universe is surprising in view of the required fine-tuning of the constants, the life-friendliness of Earth is at least prima facie surprising in view of the required fine-tuning of the factors mentioned.[3]

---

[2] See Ward and Brownlee [2000] for a defense of the view that Earth's apparent fine-tuning for life is dramatic. More recent research suggests, however, that the proportion of life-friendly planets may actually be significantly higher than previously assumed [Loeb et al., 2016]. The problem of the fine-tuned planet appears in one of the early paper on the anthropic principles [Carter, 1983], and it is discussed as a candidate analogy to the problem of the fine-tuned constants in [Manson and Thrush, 2003, p. 73] and [Greene, 2011, pp. 169f.].

[3] A potentially relevant difference is that the fine-tuning of the constants is a fine-tuning *within* the laws of nature (since the constants appear in the laws), whereas Earth's fine-tuning is a merely local affair. One could highlight this difference in an attack against the fine-tuning argument for the multiverse, perhaps along the

Unlike the apparent fine-tuning of the constants, Earth's apparent fine-tuning for life is not widely perceived as a profound puzzle. There does not seem to be any research activity that is directed, for example, at constructing a theory of planet formation according to which planets – or, more realistically, planets of a certain type, exemplified by Earth – are, in general, life friendly, as a consequence of the physical laws, which would make Earth's life-friendliness entirely expectable. The most straightforward reason as to why we do not feel that such a theory would be helpful is that, when assessing the evidential relevance of Earth's apparent fine-tuning, we seem to intuitively take into account an observation selection effect similar to the one invoked by proponents of the fine-tuning argument for the multiverse: it has been known for a long time that there are other planets beside Earth – enormously many, according to relatively recent discoveries of extrasolar planets; and, given the enormous size of our universe, it is only to be expected that at least some of the many planets in our universe are life friendly; finally, that we live on one of the (comparatively rare) life-friendly ones is unsurprising since we could not have possibly found ourselves on any of the others (nor anywhere else in life-hostile interplanetary space, for that matter).

According to this line of thought, Earth's life-friendliness can be elegantly explained by appeal to the long-suspected, now established, existence of an enormous number of extrasolar planets, many of them not life friendly, in combination with an observation selection effect. Call this perspective on Earth's apparent fine-tuning the *many planets* response to planetary fine-tuning. According to it, if we *lacked* all observational evidence for the existence of other planets besides Earth, it would be rational for us to infer that there are likely many other planets besides Earth, many of them inhospitable to life. The *many planets* response is closely analogous to the *many universes* response to apparent cosmic fine-tuning, as the rows "Our universe" and "Our planet (Earth)" in Table 4.1 indicate in a side-by-side exposition. The fine-tuning problems "Our ancestors" and "Our human civilization," outlined in the third and fourth rows, are discussed in the following sections.

In analogy to the *this universe* objection against the fine-tuning argument for the multiverse, one can set up a *this planet* objection against the *many planets* response to planetary fine-tuning. The *this planet* objection contends that the existence of many other planets does nothing to explain why Earth is life friendly. In analogy to the *this universe* response, it insists that we should not reason as if we could have existed on a different planet. According to it, if there were no independent evidence

---

lines pursued in [Colyvan et al., 2005], arguing that it does not make sense to demand an explanation for the fine-tuned constants since their values could not have been different as a matter of physical necessity. This attack, however, whether successful or not, is unrelated to the inverse gambler's fallacy charge and therefore ignored in what follows.

Table 4.1 *Four types of apparent fine-tuning*

	Type of apparent fine-tuning	Many ... response
Our universe	Values of the constants and cosmic boundary conditions right for life (*cosmic fine-tuning*)	There are multiple universes, most with the wrong constants and wrong boundary conditions for life, a few, including ours, with the right ones. As observers, we had to find ourselves in a universe with the right constants.
Our planet (Earth)	Earth's size and age, distance from central star, abundance of chemical elements, etc., right for life (*planetary fine-tuning*)	There are multiple planets, most with the wrong size and age, distance from central star, abundance of chemical elements, etc., and a few, including Earth, with the right ones. As observers, we had to find ourselves on a planet with the right conditions.
Our ancestors (over the last 500 million years)	Highly adapted to perennially changing environmental conditions, competitive in continuous struggle for survival and reproductive opportunities, capable to raise infants, etc. (*organismic fine-tuning*)	There were multiple siblings (and cousins, etc.) of our ancestors, many of them less well adapted to their environments than our ancestors, less competitive in continuous struggle for survival and reproductive opportunities, less capable to raise infants, etc. Evidently, our ancestors were among the reproductively successful organism. It is only to be expected that, as such, they were among the particularly well-adapted ones.
Our human civilization	Has so far survived despite having access to weapons of mass destruction whose use in an escalating conflict might lead to human extinction	There are multiple civilizations in different solar systems, many of them less lucky in avoiding war with weapons of mass destruction.

for many other planets, inferring their existence from Earth's fine-tuning for life would mean comitting the inverse gambler's fallacy. Manson and Thrush briefly consider the possibility that one might endorse this objection and dismiss it:

[A]ccounts that appeal to the vast number of planets in our universe (and hence the vast number of chances for conditions to be just right) surely are not to be faulted for failing to explain why *this* planet is the fit one. Clearly the "This Planet" objection (TP) is no good[.] [Manson and Thrush, 2003, p. 73]

However, there is a simple move that, at least in principle, allows its proponents to defend the *this planet* objection: they can claim that Earth's life-friendliness may just be a primitive lucky coincidence that we have to accept as such. It is naturally combined with a view, along the lines of those defended by Gould [1983] and Carlson and Olsson [1998], as discussed in Section 2.2.1, according to which life-friendly parameters are not, in any relevant sense, improbable at all. According to this *primitive coincidence* response to Earth's apparent fine-tuning in combination with the *this planet* objection, the existence of many other planets does not make it more likely that our planet is life friendly and, so, contributes nothing to the explanation of that finding. And, according to this line of thought, in the absence of independent observational evidence for other planets, the inference from Earth's fine-tuning to their existence would commit the inverse gambler's fallacy, just like the inference from fine-tuned constants to a multiverse.

Interestingly, in our actual epistemic situation, where we do have independent empirical evidence for many other planets besides Earth, it does not matter much for scientific practice whether one accepts the *many planets* or the *primitive coincidence* response to Earth's fine-tuning: proponents of both responses can agree that attempts to explain Earth's life-friendliness by appeal to the laws of physics or by appeal to some divine designer are neither needed nor promising any more. From the perspective of the *many planets* response, the existence of many other planets besides Earth already provides a satisfactory explanation of why Earth is life friendly. From the perspective of the *primitive coincidence* response, the existence of many other, mostly life-hostile planets is also relevant, though in a very different way – namely, in that it shows that planets in general are just not life friendly, which in turn indicates that no law-based or designer-based explanation of why Earth, qua being a planet, *had to be* life friendly will succeed.

4.5 The Fine-Tuned Ancestors

Consider our ancestors over the last 500 million years and picture their course of evolution across generations.[4] Now, relying on your awareness of the dangers that threaten animals in the wild and the challenges to their successful reproduction, estimate what the odds were for the members of such a large class of organisms to survive (without exception) into reproductive age, reproduce, and successfully raise at least some of their young: no doubt exceedingly low! Our ancestors must have been extremely well adapted in order to overcome all those permanent threats to their survival and reproductive success.

---

[4] The biological layperson interested in performing this exercise in imagination may profit from consulting Dawkins [2004] for a lively account of our extended evolutionary history, which focuses in particular on what is known about our common ancestors with other extant species.

A key component of the correct response to this apparent *organismic fine-tuning* of our ancestors (see the third row in Table 4.1) is the standard – and, no doubt, appropriate – Darwinian account of *natural selection* as the main mechanism of why all organisms, not only our ancestors, were and are apparently fine-tuned in the sense of being highly adapted (while other factors beside natural selection – notably, genetic drift, mutation, and migration – also play crucial roles in evolution). At any stage of evolution, organisms that are better adapted generally have better chances to survive and reproduce. As a result, organisms continue being adapted over generations, even when selection pressures vary over time.

Note that, in this appeal to natural selection to account for our ancestors' apparent fine-tuning, we have to include their same-species companions in the picture – e.g., their siblings and cousins, notably those who either did not survive into reproductive age or did so but failed to reproduce (or became the ancestors only of nonhuman organisms). Natural selection requires almost permanent "overproduction" of organisms in order to not lead into terminal population decline, so in that sense, this reply to fine-tuning invokes "many organisms."

Clearly though, despite its appeal to "many organisms" in this sense, this response to our ancestors' apparent fine-tuning differs fundamentally from *many planets* and *many universes*: while the latter are centered around appeal to an observation selection effect, the response to our ancestors' apparent fine-tuning just sketched is based on an appeal to natural selection. However, if we try to account for our ancestors' *full* apparent fine-tuning, it turns out that the appeal to natural selection does not suffice: we must either add an appeal to an observation selection effect or invoke sheer luck. Interestingly, Smolin [2007] offers a version of the *many universes* response to cosmic fine-tuning that applies natural selection at the cosmic level. His proposal is speculative, though, and does not seem to be favored by the available evidence.

That we must add an appeal to an observation selection effect or invoke sheer luck in addition to invoking natural selection can be seen as follows: in view of the theory of natural selection itself, it is reasonable to expect that those organisms that survive into reproductive age and actually reproduce, when compared to their same-species contemporaries, are, in general, particularly well adapted to the dominant selection pressures of the day. Their "degree of apparent fine-tuning," inasmuch as such a construct can be defined, is typically above average. As a consequence of their reproductive success, their phenotypic traits correlate more strongly with the phenotypic traits of next-generation organisms than those of their contemporaries with less reproductive success. Thus, at each evolutionary stage, the successfully reproducing organisms, unlike their contemporaries, seem to correctly "anticipate" the subsequent turns of evolution.

Evidently, all our ancestors were so lucky to survive into reproductive age and reproduce, so most of them were probably particularly well adapted and correctly "anticipated" subsequent turns of evolution. But why were they so lucky or, equivalently, why were *we* so lucky that none of our ancestors failed to survive into reproductive age and to reproduce?

There are two coherent ways to respond to this question. The first, which – in analogy to *many universes* and *many planets* – may be called *many organisms*, invokes an observation selection effect: our ancestors are not "randomly chosen" organisms in the evolutionary history of our species. We focused on them by using a criterion – being our ancestors – which entails survival into reproductive age and successful reproduction. It is only to be expected that organisms that conform to this criterion are, on average, as adapted ("fine-tuned") as reproductively successful organisms usually are.

But there is a second coherent way to respond – namely, to refuse giving any explanation of why our ancestors were so particularly well adapted beyond citing sheer luck. Evidently, this reaction parallels the *primitive coincidence* response to Earth's apparent fine-tuning encountered in the previous section. The latter reaction seems coherent as well.

The ideological gulf between both responses may seem deep. Notably, those who opt for the *primitive coincidence* response may claim that those who adopt *many organisms* commit the inverse gambler's fallacy. On the level of scientific practice, however, there seem to be few significant differences, which is due to the fact that there are very strong independent reasons to believe that our ancestors had many same-species contemporaries, many of whom were less well adapted to the dominant environmental selection pressures of the day than our ancestors were. Given this shared belief, all parties agree that it is unpromising to try to explain why precisely our ancestors – rather than some of their reproductively less lucky contemporaries, say – were so well adapted that their chances to survive and reproduce were comparatively high. A single scientific individual may switch forth and back between both perspectives – "primitive coincidence" versus "observation selection effect" – without displaying any irritating or incoherent behavior in practice. To conclude, because we have strong independent reasons to believe that our ancestors had multiple siblings who were less lucky, reflecting on them, while illuminating and instructive in itself, has not given us any novel reason to believe that we can determinately assess whether reasoning that has the same form as the fine-tuning argument for the multiverse is fallacious or not.

4.6 The Lucky Civilization

In a variation of the previous two examples, we can consider hypothetical alternative past fates of human *civilization*. Humanity has faced – and still faces, according to some, more urgently than ever – various potentially catastrophic risks, some of which having the potential to threaten its continued existence [Rees, 2003; Ćirković et al., 2010; Torres, 2017].

For example, the discovery of nuclear fission chain reactions created the possibility of an all-out nuclear war between superpowers. Such a war may, in turn, lead to dramatic climate change in the form of "nuclear winter," which may threaten humanity's very existence. In the future, risks posed by the dangers of greenhouse gas–induced climate change, biological weapons, nanotechnology, or hostile super-intelligence might be even greater.

When one reflects on those existential threats to humanity's survival and considers them to be serious, one may become puzzled why we still exist. One possible reaction to this puzzlement could be to wonder whether the risks of, say, nuclear or biological weapons being deployed on a grand scale might, in fact, be lower than one might naively have assumed. Coming to the conclusion that these risks are, indeed, much smaller than initially thought would be somewhat akin to having found, say, a mechanistic explanation of why our universe and/or our planet *had* to be life friendly even though this initially seemed to require an extraordinary coincidence.

But there is also a possible reaction to the puzzlement of why we are still there despite all those existential risks which is structurally similar to the *many universes* response to cosmic fine-tuning. It goes as follows: perhaps the existential risks are at least as great, perhaps even greater, than widely assumed, perhaps even so large that most civilizations die out quickly after developing the technologies that give rise to them. Even that would not turn our continued existence into a miracle, provided that there are, in fact, millions (or even billions or trillions) of other civilizations of intelligent beings out there on life-friendly planets within other solar systems of our universe.

For, if there are such civilizations, it is unsurprising that at least a few of them survive some decades past the technological breakthroughs that give rise to the most serious existential risks threatening them. Trivially, we could not possibly have found ourselves to be members of a civilization that has already died out. In the light of this observation selection effect, the assumption that there are multiple civilizations with at least some surviving for a while after the first appearance of the most serious risks is capable of "explaining" why our civilization still exists despite severe extinction risks. This "explanation" has the same structure as the "explanation" in terms of a multiverse hypothesis of why we exist despite the required cosmic fine-tuning. The *many civilizations* response to the question of why we survived at least some decades of exposure to various existential risks is thus analogous to the *many universes* response to cosmic fine-tuning, the *many planets* responses to planetary fine-tuning and the *many ancestors* response to organismic fine-tuning.

How convincing is this *many civilizations* response to the lucky civilization problem? Can we draw a lesson from it concerning whether we should accept the *many universes* response to our universe's fine-tuning for life – i.e., the standard fine-tuning argument for the multiverse?

It would not be surprised if, in the case of the lucky civilization problem – unlike in the case of, say, the fine-tuned planet problem – many readers will intuitively prefer the *primitive coincidence* response.[5] Such readers will have sympathy for the view that the hypothetical existence of many other civilizations that became exposed to the same risks as we did and predominantly succumbed to them very soon would not make it any less remarkable or mysterious why we still exist despite becoming exposed to those same risks. We have simply been somewhat lucky – perhaps very lucky! – until now to have evaded extinction so far. We cannot infer anything concerning the supposed (past, present, or future) existence of those other civilizations from the observation that we still exist despite facing various existential risks.

I suspect that many readers will find this *primitive coincidence* response to the lucky civilization problem appealing – even readers who are sympathetic to the *many planets* response to Earth's fine-tuning. While I do not see any good systematic reasons for assuming that the two problems are disanalogous, there may be strong *emotional* reasons for treating them differently, for the following reasons.

One interesting aspect of the *many civilizations* response to the problem of the lucky civilization is that it has unpleasant consequences for what we should expect concerning humanity's future fate: if we accept it, assume that there are many other civilizations facing the same or similar existential risks as we do, and further assume that most of those civilizations succumb to these threats rather soon when they arise, it is still not surprising why we exist despite being threatened by those same risks. For, given the overall large number of civilizations, it is unsurprising that there are at least some civilizations that survive for some decades, centuries, or even millenia, and we could not possibly have found ourselves to be members of a civilization that has died out. Our existence until now, despite facing various existential risks, is thus no good reason to expect having similar luck in the future – at least not if we accept the perspective of the *many civilizations* response.

As we see, there is a somewhat gloomy line of thought that follows from endorsing the *many civilizations* response to our survival until now. People may shy

---

[5] I may be wrong with this guess. Intuitions concerning which responses to the fine-tuning problems are adequate differ sharply between people. I have discussed the inverse gambler's fallacy charge against the other three fine-tuning arguments (for many universes, many planets and many ancestors) with several audiences and colleagues. Some find it obvious that the "many" responses commit the inverse gambler's fallacy and are obviously flawed, whereas others perceive the inverse gambler's fallacy charge as obviously misguided and the inferences to the "many" responses as fine. In my experience, these intuitions not only differ strongly between people; they often significantly develop over time within the same people.

away from having to commit themselves to it, and this may make them reluctant to embrace the *many civilizations* response. Proponents of the *primitive coincidence* response to our survival, in contrast, regard the hypothetical existence of other civilizations as irrelevant. They are, thus, not committed to such gloomy reasoning. For them, it seems natural to hypothesize no more luck than necessary, which means that they can reasonably infer from our existence despite various existential risks in the past that the existential risks mentioned earlier (alongside various others) may not be not quite as large as sometimes thought. This could account for why people might not *want* to embrace the *many civilizations* response even if they embrace at least some of the other "many" responses: its implications, even if only dimly perceived and not explicitly derived, may just be too scary.

One could also try to defend the view that there is a profound difference between the *many civilizations* problem and, say, the *many planets* problem by arguing that it is somehow more conceivable that we could have lived on a different planet than that we could have been part of a different civilization. But I do not expect such an argument to be ultimately compelling. There does not seem to be a clear-cut systematic difference between the different problems, such that the "many" responses are adequate for the fine-tuned universe, the fine-tuned planet, and the fine-tuned ancestors, whereas the *primitive coincidence* response is adequate in the problem of the lucky civilization.

The *many civilizations* response and the *primitive coincidence* response to the lucky civilization problem differ in whether they atttribute any significant consequences to the observation selection effect that arises from the fact that we could not have found ourselves to be members of a civilization that has already died out. But both are compatible with the importance of another observation selection effect, which arises from the fact that certain catastrophes must have been absent from Earth in its more recent history for human civilization to arise in the first place.

To understand the importance of this latter observation selection effect, it is useful to realize that we should not be surprised that, for example, no very large asteroid has hit Earth in the past few million years, even if large asteroid impacts are actually very common. For a large asteroid impact might have prevented our civilization from ever developing, and against the background of our own existence as a civilization, its absence in Earth's more recent past should not tempt us to assign such impacts an overall low probability, notably not a low future probability. Ćirković et al. [2010] dub this phenomenon the "anthropic shadow" effect and draw general lessons from it for the assessment of future human extinction risks. As they persuasively argue, any civilization, including our own, will have a tendency to systematically underestimate the existential risks that it faces when extrapolating from records about times before it came into existence.

Proponents of the *primitive coincidence* response to the lucky civilization problem do not have to deny the importance of the observation selection effect associated with the anthropic shadow when assessing future extinction risks. Where they differ from proponents of the *many civilizations* response is concerning the importance of the observation selection effect related to risks that our civilization has faced since it started to exist.

To conclude, all in all, proponents of the *many civilizations* response will be more pessimistic regarding humanity's future prospects than proponents of the *primitive coincidence* response. But all should be wary of extrapolating from past records the probabilities of risks that would have prevented humanity from existing in the first place and should take care to do justice to the anthropic shadow effect.

4.7 A Glance at the Fermi Paradox

The Fermi paradox arises from the tension between the observation that we have not detected any signs of extraterrestrial intelligence yet and estimations according to which intelligent forms of life are quite common in our galaxy.[6] Interestingly, the fact that we have not found any signs of extraterrestrial intelligence yet imposes constraints on viable responses to the lucky civilization problem. Notably, proponents of the *many civilizations* response cannot credibly hypothetize an arbitrarily large number of civilizations in our cosmic neighborhood without getting into conflict with the it.

The impact of the Fermi paradox on the *primitive coincidence* response to the fine-tuned planet and the *primitive coincidence* response to the lucky civilization is a bit more complicated. While there are strong systematic parallels between these responses, the Fermi paradox somewhat limits the options for combining them.

The key insight to understanding those limits is that unless proponents of the *primitive coincidence* response to the fine-tuned planet problem are happy to accept the existence of life on Earth as an *extreme* coincidence, they will have to assume that life is at least somewhat abundant in our universe. Similarly, unless proponents of the *primitive coincidence* response to the lucky civilization problem are happy to regard our survival until now as an *extreme* coincidence, they will have to assume that it is at least somewhat common for civilizations to survive beyond the development of risky technologies.

---

[6] Whether the Fermi paradox really deserves to be regarded as a "paradox" is controversial. A careful recent investigation [Sandberg et al., 2018] takes into account all the uncertainty in our knowledge of factors that enter into estimates of the abundance of intelligence in our universe. It comes to the conclusion that, as far as current scientific knowledge goes, any reasonable prior probability of there being *no* other intelligent life in our galaxy is quite substantial. According to this result, the Fermi paradox does not, in fact, deserve the label "paradox."

If one adds to this package of views the (debatable) assumption that it is at least somewhat common for life on a planet to result in a civilization,[7] one arrives at the conclusion that civilizations that exist for roughly as long as we have already existed are at least somewhat common (for, otherwise, one would have to assume that we have been lucky to a highly atypical degree). Depending on what "somewhat common" then turns out to mean in quantitative terms, this conclusion may be ruled out by the Fermi paradox, which sets an upper boundary on the abundance of intelligent life in our universe.

It is worthwhile to note that the *many ...* responses as well as the *primitive coincidence* responses to the problems discussed here are normative and concern the structure of rational thought. They are not descriptive assumptions about the world. Thus, their respective tensions with the Fermi paradox do not in any way speak against or in favor of any of them them, not even partially. All they do is impose constraints on viable ways to combine them.

An interesting observation, highlighted by Bostrom [2008], applies independently of which combination of responses we endorse: if we hope for a long future of our civilization and assume that its chances of continued survival are similar to those of hypothetical civilizations on other planets, we should hope that life-friendly planets are rare. For, if life-friendly planets are abundant, civilizations are likely to appear much more often. But, assuming that civilizations are, in general, detectable from the distance, this is compatible with the Fermi paradox only if such civilizations overwhelmingly exist only for a short time and get annihilated rather quickly. Thus, if we hope that the odds of a long future for humanity are quite good, we should hope that life-friendly planets are a rare exception. The proponent of the *primitive coincidence* response to planetary fine-tuning may therefore hope that the coincidence she/he invokes is extreme, even if she/he rationally assumes that it is not.

4.8 Back to the Multiverse

Apart from the fact that natural selection plays a crucial role in the problem of our fine-tuned ancestors, all the three problems discussed in the previous sections seem relevantly analogous to the problem of the fine-tuned universe. But in the absence of debate concerning what we can rationally infer from the fine-tuning of our planet and the fine-tuning of our ancestors may well simply be a consequence of the fact that we have independent empirical evidence for other planets and for multiple sibling organisms of our ancestors. The absence of debate concerning whether we

---

[7] This assumption can be dropped if one focuses on "life in a civilization" rather than "life" in the problem of the fine-tuned planet from the start.

can infer from our survival as a civilization until now that there are likely many other civilizations, many of them less lucky, may simply reflect failure to appreciate the parallel and/or uneasiness concerning the likelihood of our own extention.

In any case, the absence of debate does therefore not strongly suggest a stable expert consensus that the inference from fine-tuning to the respective *many* response is rationally acceptable. Indeed, in all cases, it seems coherent to level the inverse gambler's fallacy charge against the inference from our existence, despite the odds, to the respective *many* response. Thus, despite the quality of the analogies, considerations on our fine-tuned planet, our fine-tuned ancestors, and our lucky civilizations unfortunately do not seem to be able to help us reach a verdict on whether the inference from our universe's fine-tuning for life to many other universes is fallacious or not.

Whereas arguments based on urn and casino scenarios suffer from the fact that one can doubt whether those scenarios are really relevantly analogous, the fine-tuned planet and fine-tuned ancestors problems are ultimately of little help because, were it not for the existence of independent evidence for many planets and many organisms, the dialectical situation with respect to them would *exactly* parallel the dialectical situation with respect to the fine-tuned constants.

The difficulties may well be principled: perhaps the problem of the fine-tuned universe is just too different from *any* epistemic problem for which fairly uncontroversial standards of rationality dictate a unique solution. We should take the possibility seriously that the question of whether the inference from fine-tuning for life to multiple universes commits the inverse gambler's fallacy simply has no determinate answer, at least not in the light of established standards of rationality. The fact that the debate about the standard fine-tuning argument for the multiverse and the inverse gambler's fallacy has been around for three decades, resulting in little more than hardening lines, may be taken to indicate that this is indeed the case.

What would our epistemic situation be like if we had convincing independent empirical evidence for many other universes with different constants (or convincing independent evidence that the constants differ across space-time in our own universe)? We would then be in a situation with respect to the fine-tuning of the constants that is similar to our actual situation with respect to the Earth's and our ancestors' fine-tuning: we could either regard the life-friendliness of the constants as elegantly explained by the (independently established) existence of the other universes in combination with an observation selection effect; or we could regard the project of explaining why the constants have the values that they have where we are as obsolete because there would evidently be no principled reasons as to why they are what they are where we are.

Notably, if we had independent evidence for other universes with different constants, the appeal to a divine designer to explain why our own universe is life friendly would lose its appeal. As conceded by White, "while we might suppose that a designer would create some intelligent life somewhere, there is little reason to suppose it would be here rather than in one of the many [hypothesized] other universes" [White, 2000, pp. 273ff.].

Can we hope to obtain independent empirical support for the existence of other universes *if* such universes exist? This question will be tackled systematically in Chapters 7 and 8. The moral of the present chapter is that the fine-tuning considerations – assuming they are accurate and the life-friendliness of our universe is indeed highly sensitive to variations in various parameters – do not dictate assigning a high degree of belief to the general multiverse hypothesis: it seems coherently possible, though perhaps not rationally inescapable, to accept the life-friendliness of our universe as a primitive coincidence. To the extent that this coincidence is remarkable, one would not regard it as rendered any less remarkable by the existence of multiple other universes where the parameters have different values.

5

Problems with Priors

This chapter continues the discussion of the standard fine-tuning argument for the multiverse, switching to the language of Bayesian (subjective) probabilities, in which the argument is presented in Section 5.1. After highlighting the desideratum of motivating a non-negligible (ur-) prior for the multiverse in Section 5.2, I assess a worry due to Cory Juhl [2007] (in Section 5.3) about belief in the multiverse as based on the standard fine-tuning argument for the multiverse: that, even if the inverse gambler's fallacy charge could be rebutted, such belief would inevitably rely on illegitimate *double-counting* of the fine-tuning evidence and would, hence, be irrational.

In Section 5.4, I argue that this concern can be assuaged, at least in principle: Juhl's qualms notwithstanding, it is coherently possible that there be empirical evidence in favor of some specific multiverse theory – and thereby, derivatively, for the generalized multiverse hypothesis – whose evidential impact is independent of the fine-tuning considerations.

An additional benefit of the probabilistic formalism, discussed in Section 5.5, is that it can be used to clarify why it is so difficult to determine whether the standard fine-tuning argument for the multiverse is fallacious or not: the difficulty can be linked to an ambiguity in the background knowledge based on which the impact of the finding that the conditions are right for life in our universe is assessed.

5.1 Bayesian Formulation

Like the fine-tuning argument for divine design, the standard argument from fine-tuning for the multiverse is often presented in Bayesian terms, using subjective probability assignments. In a simple exposition, probability assignments, aimed at being reasonable, are compared for a general single-universe theory T_U and a multiverse theory T_M. One can, in this context, think of T_U as a disjunction of specific single-universe theories T_U^λ that differ on the value of some parameter λ

that collectively describes the (by assumption, uniform) laws and constants in the one universe that exists. The multiverse theory T_M should be thought of as entailing that the parameters (laws, constants, and boundary conditions) are different in the different universes that exist according to T_M and scan over a wide range of values.

As the evidence whose impact is assessed, the standard argument uses the proposition R that there is (at least) one universe with the right parameters for life. As discussed in the previous chapter, the inverse gambler's fallacy charge states that the argument collapses if R is replaced by H: that the parameters are right for life *here*, in *this* universe. The impact of switching from considering R to considering H is discussed in Section 5.5. For the first four sections of this chapter, I assume, for the sake of the argument, that the inverse gambler's fallacy charge can be successfully addressed, and I focus on another potential difficulty for the standard fine-tuning argument for the multiverse: Juhl's double-counting charge.

For the sake of perspicuity, it will be useful to explicitly include our background knowledge B_0 in all formulas. The prior probabilities assigned to T_M and T_U before taking into account R are written as $P(T_M|B_0)$ and $P(T_U|B_0)$. The posterior probabilities $P^+(T_M)$ and $P^+(T_U)$, in turn, are supposed to reflect the rational credences of an agent who not only knows that a life-friendly universe requires fine-tuning but also knows R – i.e., that there is, indeed, a life-friendly universe.

Updating the probability assignments from the prior to the posterior probabilities is done in accordance with Bayesian conditioning such that, for the ratio of these posteriors, we obtain

$$\frac{P^+(T_M)}{P^+(T_U)} = \frac{P(T_M|R, B_0)}{P(T_U|R, B_0)} = \frac{P(R|T_M, B_0)}{P(R|T_U, B_0)} \frac{P(T_M|B_0)}{P(T_U|B_0)}. \tag{5.1}$$

If the multiverse according to T_M is sufficiently vast and varied, life appears in it with high likelihood, so the conditional probability $P(R|T_M, B_0)$ is very close to one:

$$P(R|T_M, B_0) \approx 1. \tag{5.2}$$

If we assume that, on the assumption that there is only a single universe, it is improbable that it has the right conditions for life, the conditional prior $P(R|T_U, B_0)$ will be much smaller than one:

$$P(R|T_U, B_0) \ll 1. \tag{5.3}$$

Eq. (5.3) is supposed to be the probabilistic implementation of the arguments according to which life requires fine-tuned laws and constants.

Eqs. (5.2) and (5.3) together yield $P(R|T_M, B_0) \gg P(R|T_U, B_0)$, which entails $P(R|T_M, B_0)/P(R|T_U, B_0) \gg 1$ – i.e., a ratio of posteriors that is much larger than the ratio of the priors:

$$\frac{P^+(T_M)}{P^+(T_U)} = \frac{P(T_M|R, B_0)}{P(T_U|R, B_0)} \gg \frac{P(T_M|B_0)}{P(T_U|B_0)}. \tag{5.4}$$

So, unless we have very strong prior reasons to dramatically prefer a single universe over the multiverse, i.e., unless

$$P(T_U|B_0) \gg P(T_M|B_0), \tag{5.5}$$

the ratio $P^+(T_M)/P^+(T_U)$ of the posteriors will be much larger than one and we can be rather confident that there is a multiverse.

5.2 Motivating an Ur-Prior for the Multiverse?

Rational belief in a multiverse as based on the fine-tuning argument for the multiverse requires having a solid motivation for assigning a non-negligible prior $P(T_M|B_0)$ to the multiverse hypothesis. Coming up with such a motivation is not at all straightforward because, as discussed in Section 3.3 in the context of the fine-tuning argument for divine design, that prior can only make sense as an "ur-prior," assigned from the strange perspective of an agent who is temporarily unaware of the fact that the conditions are right for life. In other words, the background knowledge B_0 is not allowed to entail any information that includes R.

To appreciate why it is important to have a well-motivated ur-prior $P(T_M|B_0)$, this difficulty notwithstanding, it is useful to consider again the *special design* hypothesis SD, which states that our universe is the product of a very powerful and intelligent creator fixated on creating a universe *exactly like ours* (i.e., with physical, chemical, biological, geographical, social, psychological, and further details all precisely as in our universe). Given this design of *special design*, our conditional credence $P(R|SD, B_0)$ should be one. A hypothetical "fine-tuning argument for special design," analogous to the fine-tuning argument for the multiverse outlined earlier, would seem to give us a very strong case for SD. The reason why we may nonetheless not believe in SD is that, as said in Section 3.4, it seems ad hoc, contrived, and devoid of an independent motivation.[1] Formally, this may be expressed by saying that any reasonable hypothetical prior $P(SD|B_0)$ will reasonably be minuscule. In fact, it may well be so tiny that even after taking into account all our "confirming" evidence for SD, notably the evidence R, our posterior $P^+(SD)$ will still be negligible.

---

[1] For this, reason we would also be reluctant to accept SD by inference to the best explanation. The problem of whether the multiverse hypothesis can be motivated independently of the fine-tuning considerations arises not only on the Bayesian formulation.

Perhaps the multiverse hypothesis T_M is similarly contrived and devoid of an independent motivation as the special design hypothesis! For example, any empirically viable specific multiverse may require fine-tuning of its basic parameters to be compatible with life, no less than any viable single-universe theory. Lacking a compelling argument to the contrary, even if the finding that the parameters are right for life does turn out to confirm the multiverse hypothesis over a rival single-universe hypothesis (i.e., even if the inverse gambler's fallacy charge can be overcome), its outcome may not be rational confidence in the multiverse but only some degree of belief that, though larger than any rational hypothetical prior $P(T_M|B_0)$, is still negligibly small. The example of SD suggests that proponents of the fine-tuning argument for the multiverse should have a good case for a non-negligible hypothetical prior $P(T_M|B_0)$.

According to Juhl, however, there are principled reasons for believing that there can be no such case:

One worry that I have about an argument for an actual multiverse is that such an argument would have to somehow convince us that the prior probability $P(M)$ of a multiverse hypothesis is high enough on any reasonable probability assignment to warrant assigning it a high posterior (higher than $1/2$, say). I do not foresee a convincing a priori argument for this, either on the basis of principles of indifference, or on the basis of a rationally compelling "logical" approach to assignments of priors. But if we try to assign rational "prior" probabilities on an a posteriori basis of empirical evidence, as a means of avoiding "purely a priori" approaches and their attendant difficulties, then we might find it difficult to avoid a problem to be described below.... The problem might be called a fallacy of (evidential) "double dipping." [Juhl 2007, p. 554.]

What would be an "a priori argument" for a non-negligible (hypothetical) prior $P(T_M|B_0)$? Here, Juhl might have in mind something like Max Tegmark's considerations in favor of the *mathematical multiverse hypothesis*, according to which all mathematical structures are realized physically as universes. Tegmark's suggestion will be dealt with critically in Section 10.3. For the moment, let us focus on attempts to obtain a well-motivated $P(T_M|B_0)$ based on, as Juhl calls it, "an a posteriori basis of empirical evidence." Juhl is pessimistic about such attempts as well, or, more precisely, he is skeptical that empirical evidence for some multiverse theory could, in practice, be able to support it *independently of* the observation that the parameters in our universe are right for life.

5.3 The Double-Counting Charge Stated

It seems plausible that theories such as inflationary cosmology and string theory, which are to some degree independently attractive and combined in the landscape multiverse scenario, can provide an indirect independent motivation for the multiverse hypothesis, over and above the fine-tuning considerations. Juhl regards it as unlikely, though, that such considerations could ever help motivate a useful non-negligible (ur-) prior $P(T_M|B_0)$, due to a difficulty related to double-counting ("double dipping," as he calls it). Double-counting, to use an example by Juhl himself [Juhl, 2007, pp. 554f.], occurs when a prosecutor in court first appeals to the fact that some person's fingerprints are on the murder weapon to argue that we should seriously consider the possibility that she might be the murderer and then argues that, in view of the fact that her fingerprints are on the murder weapon, we should now become more confident that she really is the murderer. It is uncontroversial that such double-counting is fallacious.

According to Juhl, double-counting is almost unavoidably committed by any attempt to run the fine-tuning argument for the multiverse in combination with purportedly independent empirical support for some multiverse theory that might provide some specific an account of how universes with different values of the parameters are generated:

I do not foresee a plausible scenario in which what I call the "prior" of the existence of the law or mechanism for universe production can be rationally constrained to be reasonably high, without that "prior" itself depending on evidence entailing the existence of a universe much like ours. To make clear what I have in mind here, suppose that physicists come up with a "unified theory of everything' U, which entails that the existence of a multiverse has a probability of around 1/2. How will they discover such a theory? Surely U will be obtained via observations of goings-on within our universe. Such goings-on will entail the existence of our universe, that universe of which they are parts, or within which they are events. But if such events form the evidential basis for assigning a not-too-low probability to a theory M of multiverse production, then we cannot reuse or "double dip" that same evidence, the existence of our universe, to provide further support to the theory. [Juhl 2007, p. 555.]

The gist of this passage is that any observational data D that we might collect in support of some multiverse-friendly theory T_M will unavoidably entail the finding R according to which the parameters are right for life; in some universe. Juhl's argument is deceptively simple: any data D that we might collect will unavoidably be about "goings-on within our universe"; goings-on within our universe unavoidably "entail the existence of our universe"; and, since our universe has the right parameters for life, its existence entails the existence of a universe where the parameters are right for life, i.e., it entails R. And if, indeed, any D entails R, then basing one's prior $P(T_M)$ on D means implicitly using R in that step. Accordingly, using R when running the fine-tuning argument to boost our credence in T_M would mean using R *again* and would hence means illegitimate double-counting of R.

Juhl is adamant that the double-counting charge goes beyond highlighting that the fine-tuning argument in its Bayesian formulation must come to terms with the problem of old evidence. It is useful, though, to frame the double-counting charge in the language of the ur-probability solution, where it amounts to the worry that no empirical evidence whatsoever can be relevant when assessing $P(T_M|B_0)$, construed as a hypothetical prior. According to Juhl, data D that are potentially relevant to our credence in T_M – since they are from *our* universe – are likely to entail that the parameters are right for life in that universe. But this means that they cannot possibly be available to an epistemic agent in the hypothetical situation invoked by the ur-probability solution, who – by assumption – is unaware that the parameters are right for life. If Juhl is right, it is, in principle, impossible to motivate *any* chosen value for the hypothetical prior $P(T_M)$ by appeal to empirical evidence, notably any non-negligible value.

Is it plausible, if we look at the example of the landscape multiverse, that multiverse-supporting empirical data will inevitably entail R: that the parameters are right for life? Prima facie, it may seem quite realistic. The CMB data, for example, which support inflationary cosmology and thereby (indirectly) the landscape multiverse, may indeed entail at least large bits of the evidence R according to which the parameters are right for life: the CMB radiation is electromagnetic, so a detailed description of it will include information about the values of the constants associated with electromagnetism. Moreover, a detailed account of *how* an early inflationary phase – if it occurred – resulted in CMB fluctuations as detected today will inevitably rely on auxiliary assumptions concerning other fundamental interactions that have shaped our universe in the meantime. As a consequence, the constants associated with other interactions – the weak and strong nuclear interactions and gravity – are likely to be significantly constrained and, thus, in part, entailed – by such an account as well.

To sum up, Juhl's worry that empirical data that favor a multiverse cosmology will entail that the parameters are right for life (or at least significant parts of it) deserves to be taken seriously. If Juhl is correct, multiverse proponents who rely on the standard fine-tuning argument for the multiverse face a dilemma: they can either run that argument, but without relying on a well-motivated, non-negligible (hypothetical) prior $P(T_M|B_0)$ (thus, potentially ending up in a similar situation as those who run the fine-tuning argument for *special design* without having a good independent case for a non-negligible hypothetical prior $P(SD|B_0)$); or they can make a well-founded assignment of $P(T_M|B_0)$ based on observational data D (perhaps indirectly via appeal to empirical support for some multiverse-friendly theory T), but then they are no longer able to run the fine-tuning argument without counting R twice because D inevitably entails R. In the following, final, section, I explore how multiverse friends can address this challenge.

5.4 The Standard Fine-Tuning Argument for the Multiverse
without Double-Counting

Fortunately for proponents of the standard fine-tuning argument for the multiverse, it is at least in principle possible to motivate a non-negligible hypothetical prior $P(T_M)$ using empirical data D and to subsequently run the fine-tuning argument without the double-counting of R. The key idea is that, for suitable data D, there are generally parts or aspects D^* of D that do not entail R and may still be multiverse supporting.

To see this, consider the following: suppose that we have indeed managed to acquire data D that help us motivate the ascription of a non-negligible $P(M)$. For the sake of the argument, let us suppose that, as Juhl suggests, these data D entail parts or all of R. Now suppose further – without loss of generality – that D, specified in propositional form, is closed under deductive inference (alternatively, assume that D is the deductive closure of our multiverse-suggesting empirical data in propositional form). Generally, D will *not* be logically equivalent to R – even if D entails R – unless R entails all of D, for which, however, there seems to be no systematic reason. In general, there will be a nonempty subset D^* of D that is not entailed by R; i.e., it is logically compatible with $\neg R$. As I argue next, it is possible for this D^* to be multiverse supporting on its own.

To check whether D^* can legitimately be used to motivate a non-negligible hypothetical prior $P(M)$, we have to go back to the ur-probability approach to Bayesianism and consider a hypothetical agent who lacks the evidence R but has otherwise as much of our background knowledge as coherently possible. And whatever that background knowledge comprises – remember, we are considering an agent who, bizarrely, is unaware of her own existence – there seems to be no reason to doubt that it will include D^*. For, by assumption, we are aware of D^* and, by construction, D^* is not entailed by R. If it happens that D^* alone can be appealed to when motivating a non-negligible $P(M)$, then running the fine-tuning argument based on the so-motivated $P(M)$ does not involve any double-counting and may potentially result in a sizeable $P^+(M)$.

Avoiding double-counting along these lines is possible only if two conditions are met: first, the evidence R must not entail our multiverse-supporting data D (for, otherwise, D^* would be empty); second, D^* on its own – and not merely the full D – must be multiverse supporting. Taken together, these two conditions boil down to the simple (and unsurprising) requirement that, for the standard fine-tuning argument to result in rational belief in a multiverse, there must be *independent* empirical support for the multiverse over and above the finding that the parameters are right for life despite the required fine-tuning. Priors (or hypothetical priors) play no role in this requirement. This is an appealing feature because problems with double-counting are not by themselves conceptually tied to difficulties with the assignment of (hypothetical) priors, so it makes sense that we can ultimately assess the double-counting issue without considering specific (hypothetical) priors. What would really vindicate the double-counting charge is an argument as to why there cannot possibly be any independent empirical support for the multiverse over and above the finding that the parameters are right for life despite the required fine-tuning.

It seems unlikely, though, that such an argument will be forthcoming. Consider the support for inflationary cosmology by the [Planck Collaboration, 2016] data on the cosmic microwave background as an example. Now suppose – perhaps unrealistically, but not incoherently – that this support can be translated into support for the landscape multiverse scenario of which inflationary cosmology is an ingredient. Clearly, not all aspects of the CMB data are entailed by R, notably not all those that are taken to support inflation. If the aspects of the CMB reported by [Planck Collaboration, 2016] that favor inflation were all entailed by R, scientists would have been able to predict them, based on R alone and without any help from inflationary cosmology. So we have here empirical data D^* – in this case, concerning the CMB – which are part of our complete multiverse-suggesting data D but not entailed by R. Since these data D^* are confirmatory, on their own, of a multiverse-suggesting theory (inflation), they are (at least indirectly) multiverse suggesting by themselves. If we use them to help motivating a non-negligible $P(T_M|B_0)$, we can run the fine-tuning argument for the multiverse based on the so-motivated $P(T_M|B_0)$, potentially resulting in a sizeable posterior $P^+(T_M)$, without committing double-counting.

In practice, empirical data such as those on the CMB may support inflation only inasmuch as there are models of inflation according to which inflation is non-eternal and does *not* give rise to a multiverse scenario. According to Alan Guth, the founder of inflationary cosmology, everything that is not in principle forbidden by physical laws will end up happening in the eternally inflating universe, which, as Smeenk [2014] points out, threatens to make inflation self-undermining inasmuch as it leads to eternal inflation. To conclude, while there seems to be nothing wrong *in principle* with the idea of independent support for some multiverse theory over and above that provided by the standard fine-tuning argument for the multiverse (provided we accept that argument), obtaining such support may *in practice* be exceedingly difficult. This suspicion will be confirmed in the more detailed discussion of the prospects for testing specific multiverse theories in Chapters 7 and 8.

5.5 Rebutting the Inverse Gambler's Fallacy
Charge – What It Would Take

In the Bayesian framework, the central issue in the debate about the fine-tuning argument for the multiverse and the inverse gambler's fallacy is connected to whether the analog of Eq. (5.2) that is obtained when replacing R by H is plausible:

$$P(H|T_M, B_0) \approx 1, \tag{5.6}$$

where, to recapitulate, H stands for "this universe here is right for life."

Proponents of the inverse gambler's fallacy charge argue that the switch from considering $P(R|T_M, B_0)$ to considering $P(H|T_M, B_0)$ is necessary, typically by invoking a widely accepted rule of Bayesian inference – namely, the *requirement of total evidence*: that, starting with our priors, we should always condition on the totality of the evidence available to us in order to arrive at our rational credences. An example with which this is often illustrated concerns a person who comes across an animal that looks dangerous but that, as the person knows, is not actually dangerous. The person's rational actions – notably, whether to flee or not to flee – will be guided by her full available evidence: "This here is a dangerous-looking animal that is not actually dangerous," not by the partial evidence "This here is a dangerous-looking animal." Conditioning only with respect to the latter, partial, evidence might make it appear wise to flee, even if this would not at all be rationally required.

Proponents of the inverse gambler's fallacy charge argue that the shift from R to H, as dictated by the requirement of total evidence, is fatal to the fine-tuning argument for the multiverse: whereas Eq. (5.2) may be plausible, Eq. (5.6), according to them, is not; for, as extensively discussed in the last chapter, as they see it, whether there are any other places in the overall cosmos where the parameters are right for life is irrelevant to whether they are right for life here, in this universe, where we live.

Those who wish to defend the standard fine-tuning argument for the multiverse against this charge have two options: they can either deny that $P(H|T_M, B_0)$ should be considered instead of $P(R|T_M, B_0)$; or they can acknowledge that, in accordance with the requirement of total evidence, $P(H|T_M, B_0)$ should be considered, but argue that Eq. (5.6) is plausible after all.

Epstein [2017] defends the standard fine-tuning argument for the multiverse by opting for the first response. According to him, the requirement of total evidence does not apply in this context because we did not predesignate our universe as the object of our study prior to determining whether it is life friendly. As argued in the previous chapter, one may question Epstein's line of thought because neither did we receive the information that at least one universe is life friendly by means of a process that could also have resulted in a negative finding.

Bradley [2009], who also defends the standard fine-tuning argument for the multiverse, opts for the second response – at least he does not explicitly advocate rejecting the requirement of total evidence as applied to this case. But, as he argues, similarly in its consequences to Epstein's considerations, the "bias" in how we came to select our universe as the object of our study – that we could not possibly have observed a lifeless universe – impacts the result of conditioning with respect to the fact that this universe is life friendly. The general rule that is relevant here according to him, applied to The urn examples discussed in Section 4.2, is

[I]f the process [by means of which we select our object of study] is biased toward p, and an instantiation of p is successfully found, the evidence confirms hypotheses in which there is a large population rather than a small one. [Bradley, 2012, 66]

He applies this rule to the fine-tuned universe problem as follows:

[B]y what procedure has our universe, Alpha, been selected for observation? Alpha is more likely to be observed than other universes because Alpha has the right constants, and only universes with the right constants can be observed. We have successfully found a universe with the right constants by a biased method, so we end up with confirmation of Many universes. [Bradley, 2012, 66]

In effect, according to Bradley, there is little or no difference in effects between conditioning with respect to H and conditioning with respect to R – provided that, when conditioning with respect to H, we properly account for the observation selection effect that arises from the fact that we could not have found ourselves in a universe that is not life friendly.

My own view on this dispute is that there is no generally correct answer concerning with respect to which bit of evidence R or H one should condition. Rather, whether R or H should be used depends entirely on the background knowledge B_0 against which the conditioning is performed.

Specifically, consider background knowledge B_0, which leaves it completely open whether some specific universe Alpha – ours – is life friendly but entails that all the information we can possibly get concerning a universe's being life friendly will inevitably be about Alpha. In that case, Alpha was effectively selected by a procedure among universes (whether there are multiple others or not) that was not biased toward life-friendliness and was subsequently found to be life friendly. If one uses such background knowledge, one has to condition with respect to H, and such conditioning does not yield any shift in favor of a multiverse. Conditioning (only) with respect to R would amount to committing the inverse gambler's fallacy.

Now consider, in contrast, different background knowledge B_0, which does not dictate in which universe we will find ourselves – if we exist – provided there is at least one life-friendly universe at all. In that case, conditioning with respect to R gives the correct result, as does conditioning with respect to H and accounting for the observation selection effect that we could not possibly have existed in a life-hostile universe, in line with Bradley's considerations. For such background knowledge B_0, conditioning with respect to either R or H does result in a shift in favor of a multiverse.

The question of which background knowledge B_0 to use should be based on pragmatic considerations. There is no matter of fact concerning which background knowledge is the "true" one.

Which background knowledge B_0 we should preferably use *in practice* depends on for which background knowledge we can come up with reasonably motivated priors $P(T_U|B_0)$, $P(T_M|B_0)$ and conditional probabilities $P(H|T_U, B_0)$, $P(H|T_M, B_0)$. In principle, if one suitably adjusts the priors $P(T_U|B_0)$, $P(T_M|B_0)$ according to what is included in the background knowledge B_0, the final outcome of conditioning with respect to H should be the same, regardless of which B_0 is used. Our rational posterior credences in T_U and T_M should not depend on how we partitioned our knowledge in background knowledge and evidence used in conditioning.

The same applies to the other fine-tuning problems discussed in the last chapters and the opposition between the *many ...* responses and the lucky coincidence responses to them: in a Bayesian setting, both responses correspond to different ways of conditioning with respect to our (ongoing) existence; which way is correct depends on which prior is used. The responses differ in their consequences only if priors $P(T_U|B_0)$, $P(T_M|B_0)$ (where T_U and T_M are the respective "*single...*"- and "*many...*"-theories) are not adjusted for what is contained in B_0 and what is not. If those probabilities are appropriately adjusted for background knowledge used, there will not be in any difference in the posteriors assigned.

A difference in posterior credences does emerge between the *many ...* responses and the lucky coincidence responses if proponents of the former assume that, for the background knowledge B_0 that they operate with, the priors $P(T_U|B_0)$ and $P(T_M|B_0)$ are roughly equal, whereas proponents of the latter assume that, for the background knowledge B_0 that they operate with, $P(T_U|B_0)$ is significantly larger than $P(T_M|B_0)$. In that case, the posterior of T_M will come out as significantly larger than the posterior of T_U in the former case, and vice versa in the latter.

One can interpret these considerations as providing the basis for peaceful coexistence between the *many ...* responses and the *primitive coincidence* responses: ultimately, the disagreements between their proponents can be traced to disagreements about reasonable priors. And, at least from a thoroughly subjectivist Bayesian perspective, such disagreements are to be expected and do not indicate lack of rationality on anyone's part.

Inasmuch as this is the adequate moral to be drawn, it reinforces the conjecture made at the end of the last chapter: that established standards of rationality may just not suffice to decide between the *"many. . ."*-responses and the *primitive coincidence* responses. For inasmuch as there are no rationally preferred priors for the theories T_M and T_U, given more or less clearly specified background knowledge B_0, the question of what the rational posteriors are simply does not have a preferred answer.

One can try to solve this problem by identifying some specific suitable background knowledge B_0 for which one can motivate reasonable values of the priors $P(T_U|B_0)$ and $P(T_M|B_0)$ in a compelling way. But here the severe difficulties discussed previously in this chapter strike back. There we saw how difficult it is, in the light of Bayesianism's old evidence problem, to specify what should be included in B_0 such that we can motivate reasonable priors $P(T_U|B_0)$ and $P(T_M|B_0)$, given the constraint that B_0 may not give away the fact that life exists and that we exist. In view of these difficulties, I do not see a clear perspective for determining which type of background evidence we should use. Nor does it seem clear *how* the background knowledge B_0 would, in practice, actually tie our possible existence to some specific universe Alpha or fail to do so.

Finally, it should be noted that *even if* proponents of the standard fine-tuning argument for the multiverse can identify some suitable candidate background knowledge B_0 for which conditioning with respect to H leads to confirmation of the multiverse hypothesis *and* reasonable priors $P(T_U|B_0)$ and $P(T_M|B_0)$ can be motivated, this would not resolve all the serious problems that they face. For our full evidence not only includes the proposition H – that the parameters are right for life in *this* universe – but the even more specific statement $[\lambda_0]$: that they have some specific value configuration λ_0 in *this* universe *here*. Since λ_0 is patently compatible with life, $[\lambda_0]$ entails H, which means that

$$\frac{P([\lambda_0]|T_M, B_0)}{P([\lambda_0]|T_U, B_0)} = \frac{P([\lambda_0]|T_M, H, B_0)}{P([\lambda_0]|T_U, H, B_0)} \cdot \frac{P(H|T_M, B_0)}{P(H|T_U, B_0)}. \tag{5.7}$$

To defend the fine-tuning argument for the multiverse successfully against the inverse gambler's fallacy charge, one would have to provide a compelling argument that the second factor on the right-hand side of this equation is plausibly large (i.e., the multiverse is favored by the evidence $[\lambda_0]$). However, the first factor on the right-hand side is extremely hard to evaluate. Plausibly, $P([\lambda_0]|T_M, H, B_0)$ gives the probability of being in a universe with parameter configuration λ_0, *given* that one is an observer in some life-friendly universe within an overarching multiverse as described by T_M.

But what counts as an "observer" in the relevant sense? (Which alien in a universe with different laws does? Which of the animals on Earth? Do "stages" of humans count as observers in their own right, etc.?) The problem, again, is that B_0 is specified only so vaguely and is so remote from our actual background knowledge B_l, unavoidably, because it may not even entail that there are any humans (let alone that *we* are humans), that there is life on Earth, that the parameter configuration in our universe is λ_0, etc. In the following chapter, I investigate how we can assess the impact of fine-tuning for life on the single-universe versus multiverse question without abstracting as much from our actual background knowledge as required by the Bayesian formulation of the standard fine-tuning argument for the multiverse.

6

A New Fine-Tuning Argument for the Multiverse

This chapter has two combined aims. First, I point out that the standard fine-tuning argument for the multiverse, as discussed in the previous two chapters, differs crucially from paradigmatic instances of anthropic reasoning such as, notably, the Dicke [1961] and Carter [1974] accounts of coincidences between large numbers in cosmology. The key difference is that the standard fine-tuning argument for the multiverse treats the existence of forms of life as calling for a response and infers the existence of a multiverse as the best such response. Anthropic reasoning of the type championed by Dicke and Carter, in contrast, assumes the existence of forms of life as background knowledge when assessing whether the large number coincidences are to be expected given the competing theories. These are clearly different argumentative strategies, which should not be confused. An advantage of the Dicke-Carter argumentative strategy is that it is not susceptible to any inverse gambler's fallacy charge.

The second aim of this chapter is to propose a new fine-tuning argument for the multiverse, which – unlike the standard one – is structurally similar to Dicke's and Carter's accounts of large number coincidences. The new argument turns out to have the virtue of being immune to the inverse gambler's fallacy charge. As will become clear, it rests on one key assumption – namely, that the considerations according to which life requires fine-tuned laws and constants do not independently make it less attractive to believe in a multiverse.

6.1 The Argument from Fine-Tuning for the Multiverse as an Inference to an Anthropic Explanation?

The standard argument from fine-tuning for the multiverse as discussed in the previous two chapters is sometimes characterized as an *inference to the best explanation*

(IBE) – namely, an inference to the existence of a multiverse as the best explanation of why there is life despite the fine-tuning of parameters that is necessary for it. Because, moreover, the standard fine-tuning argument for the multiverse appeals to the strong (or weak – the distinction is arguably gradual) anthropic principle, it is sometimes characterized as an "anthropic explanation" (e.g., Leslie [1986; 1989, chapter 6], McMullin [1993, p. 376f., section 7], and Bostrom [2002]). It is doubtful, however, whether this characterization is adequate.

First, it is questionable whether even the most paradigmatic anthropic "explanations" qualify as genuine explanations at all [Earman, 1987, p. 309]. An example of such an "explanation," characterized as such by Carter in his seminal paper [Carter, 1974] that introduces the anthropic principles, is astrophysicist Robert Dicke's account of coincidences between large numbers in cosmology [Dicke, 1961]. A prominent example of such a coincidence is that the relative strength of electromagnetism and gravity as acting on an electron/proton pair is of roughly the same order of magnitude (namely, 10^{40}) as the age of the universe, measured in natural units of atomic physics. Impressed by this and other coincidences, Dirac [1938] stipulated that they might hold universally and as a matter of physical principle. He conjectured that the strength of gravity may decrease as the age of the universe increases, which would indeed make it possible for the coincidence to hold across all cosmic times.

Dicke [1961], criticizing Dirac, argues that standard cosmology with time-independent gravity suffices to account for the coincidence, provided that we take into account the fact that our existence is tied to the presence of main sequence stars like the sun and of various chemical elements produced in supernovae. As Dicke shows, this requirement dictates that we could only have found ourselves in that cosmic period in which the coincidence holds. Accordingly, there is no need to assume, as suggested by Dirac, that gravity varies with time to make the coincidence unsurprising. Carter [1974] and Leslie [1986] (see also Leslie [1989, chapter 6]) describe Dicke's account as an "anthropic explanation" of the coincidence that impressed Dirac, and Leslie discusses it continuously with the argument from fine-tuning for the multiverse.

Whether Dicke's demystifying account really deserves the honorific label "explanation" may seem doubtful, but, arguably, not much hinges on this terminological choice. Much more importantly, whereas Dicke's account of the coincidence uses life's existence as *background knowledge* to show that standard cosmology suffices to make the coincidence expectable, the argument from fine-tuning for the multiverse treats life's existence as requiring a theoretical response (rather than as background knowledge) and advocates the multiverse hypothesis as the best such response. In the following section we will study this difference using the Bayesian probabilistic formalism again.

6.2 Anthropic Reasoning: Dicke/Carter-Style

One can make Dicke's reasoning in his argument against Dirac more transparent by using the Bayesian formalism. To do so, let us denote by C the large number coincidence, by S the standard view according to which gravity is spatiotemporally uniform, and by A Dirac's alternative theory, according to which the coincidence holds universally and gravity varies with time. Dirac's idea is that the approximate identity of the two apparently unrelated large numbers is a surprising coincidence on the assumption that the standard theory is correct, whereas it is to be expected and not at all surprising on the assumption that the alternative theory is correct. This suggests the inequality

$$P(C|S, B) \ll P(C|A, B), \tag{6.1}$$

where B stands for the assumed scientific background knowledge. Probabilities assigned to S and A based on this background knowledge B prior to considering the impact of C will presumably privilege standard cosmology S over Dirac's alternative A:

$$P(S|B) > P(A|B). \tag{6.2}$$

By Bayes's theorem, posterior probabilities $P^+(S) = P(S|C, B)$ and $P^+(A) = P(A|C, B)$, which take into account the impact of C, can be computed. In line with Dirac's argument, due to Eq. (6.1) and in spite of Eq. (6.2), they may end up privileging Dirac's alternative theory A over standard cosmology S.

Dicke's considerations, however, reveal that focusing on Eq. (6.1) is not helpful. Plausibly, our background knowledge B will include some basic information about ourselves, including that we are forms of life whose existence depends on the presence of a mainline star in the vicinity. But, according to standard physics, mainline stars can exist only in that period of cosmic evolution where the coincidence approximately holds, so it is misleading to consider only the evidential impact of C. What comes closer to our full evidence is the proposition $[C]$: that the coincidence holds *here and now*, in our cosmic era. Dicke's reasoning can be expressed as the insight that we should use $[C]$ rather C as our evidence and that, instead of a probability distribution P that conforms to Eq. (6.1), we should use a probability distribution P' for which

$$P'([C]|S, B) \approx P'([C]|A, B). \tag{6.3}$$

According to Eq. (6.3), it no longer seems that finding the coincidence C realized provides any strong support for the alternative (varying gravity) theory A over the standard theory S. Since Eq. (6.2) plausibly carries over from $P()$ to $P'()$, we have

$$P'(S|B) > P'(A|B). \tag{6.4}$$

Combining Eqs. (6.3) and (6.4) and applying Bayes's theorem means that the posteriors will likely still privilege the standard theory over the alternative theory. This, in essence, is Dicke's conclusion, contra Dirac.

Interestingly, Carter presents an analogous argument in the context of the fine-tuned planet problem, to the conclusion that most "habitable" planets in our universe are likely lifeless [Carter, 1983]: the time scale on which life appeared on Earth is of the same order of magnitude as Earth's duration of existence. This finding suggests that either the time scale on which life appears on habitable planets is generally similar to the time scale on which planets exist or that the former is generally much larger than the latter. The first possibility corresponds to Dirac's view on the large number coincidence. Carter makes that case for the second along parallel lines as Dicke makes the case against Dirac's hypothesis:

[P]rovided one avoids making the habitual mistake of overlooking the anthropic principle, it can easily be seen that the observation that t_e [the time scale on which life has evolved] is comparable with the upper limit τ_0 [the lifetime of the sun] is just what would be expected if we adopt the alternative hypothesis that the intrinsically expected time $\bar{t}$ is much longer than τ_0: in this case self-selection ensures that ours must be one of the exceptional cases in which evolution has proceeded much faster than usual; on this basis it is to be expected that t_e should be comparable with τ_0 because there is no particular reason why we should belong to the even more exceptional cases in which evolution proceeds even more rapidly although, with the assumption that the Universe is infinite, such cases must of course exist.

Since this satisfactorily accounts for the observed order of magnitude of t_e there is no need to have ad hoc recourse to the a priori less plausible hypothesis that the magnitudes of such unrelated quantities as t_e and τ_0 should just happen to coincide.

An immediate consequence of Carter's conclusion – that t_e is probably much larger than τ_0 – is that life is probably very rare in our universe.

Carter's conclusion here parallels the one reached based on the *many planets* response to Earth's fine-tuning in Chapter 4. Proponents of the *primitive coincidence* response to Earth's fine-tuning can counter it by arguing that reasonable priors will privilege the assumption that life is *not* exceedingly rare in our universe.

The advantage of Dicke/Carter-style reasoning compared to the considerations investigated in the previous two chapters is that it uses our knowledge that our universe (and our planet) contain life as part of the background knowledge B. The standard fine-tuning argument for the multiverse, in contrast – and all the other *many* ... responses considered before – use this as the information whose evidential impact is assessed in the light of the fine-tuning considerations – i.e., *not* as background knowledge. As we saw in Chapter 4, the standard fine-tuning argument for the multiverse is potentially susceptible to the inverse gambler's fallacy charge. The following section develops a new argument from fine-tuning for life for a multiverse that parallels Dicke/Carter-style anthropic reasoning by treating the information that our universe is life friendly as background knowledge. This makes the new argument by construction immune to the inverse gambler's fallacy charge.

6.3 The New Fine-Tuning Argument Stated

6.3.1 Informal Formulation

The basic idea of the new fine-tuning argument for the multiverse is that the fine-tuning considerations contribute to a partial erosion of the main theoretical advantage that empirically adequate single-universe theories tend to have over empirically adequate multiverse theories – namely, that their empirical consequences are far more specific. In what follows, I give an exposition of the new argument.

The new argument contrasts multiverse theories and single-universe theories with respect to their abilities to account for the measured value of some numerical parameter λ. The parameter λ can be thought of as a multidimensional parameter configuration that collectively encodes relevant aspects of the laws and constants.

For the sake of simplicity, I assume that multidimensional there is only one fundamental multiverse theory T_M worth taking seriously. This T_M should be empirically adequate in the very weak sense that it is compatible with the existence of at least some (sub-) universe where the parameter λ has a value λ_0 that is consistent with our measurements in this universe. Next, I assume that, as the main rival to T_M, there is only one candidate fundamental single-universe theory $T_U^{\lambda_0}$ worth taking seriously. This should also be empirically adequate, which means that the value λ_0 that it ascribes to the parameter λ over the entire universe must be one that our measurements in this universe are consistent with.

Let us suppose now that λ is a parameter for which physicists come up with considerations F according to which its value requires fine-tuning to be compatible with life. Thus, only values very close to λ_0 lead to a life-friendly universe. Do those considerations F make it rational to increase one's credence in the multiverse theory T_M?

To answer this question, let us first consider our evidential situation *prior* to taking into account the considerations F according to which λ requires fine-tuning for life. An evidential advantage that we can expect the single-universe theory $T_U^{\lambda_0}$ to have over the multiverse T_M is that it makes a highly specific and, as it turns out, empirically adequate "prediction" of the value of λ. The multiverse theory T_M, in contrast, entails the existence of universes with many different values of λ. Observing any of those values in our own universe would have been compatible with T_M. This reflects the main methodological drawback of multiverse theories, often emphasized by their critics, that they tend to make few testable predictions, if any. Multiverse theories may have other theoretical virtues – e.g., at least by certain standards, elegance and simplicity.

Let us now turn to assessing the impact of the fine-tuning considerations F. To recall, these considerations reveal, to the degree that they apply to our chosen parameter λ, that the value of λ must be very close to λ_0 for there to be life. Thus, they imply that our background knowledge B_l – which, in analogy to Dicke/Carter-style reasoning, now *includes* the proposition that our universe is life friendly – constrains values of λ that we could have possibly found to a very narrow range around λ_0. So, effectively, what the fine-tuning considerations show is that, based on B_l alone, we could have predicted that we will find a value of λ that is (very close to) λ_0. Specifically, if there is a multiverse as entailed by T_M, all life forms in it that are capable of making measurements of λ will find values very close to λ_0.

But this means that the major advantage of $T_U^{\lambda_0}$ with respect to T_M erodes: after taking into account F, both these theories "predict" that observers will find the value of λ to be λ_0 (or, according to T_M, a value not that far from it). Since other potentially relevant theoretical virtues – e.g., $T_U^{\lambda_0}$'s and T_M's respective elegance and simplicity – are unlikely to be affected by the considerations F, as a net result of those considerations, the multiverse T_M becomes comparatively more attractive.

6.3.2 Bayesian Formulation

One can express these considerations in Bayesian terms, denoting by "$[\lambda_0]$" the proposition "The value of λ is λ_0 in our own universe." If $T_U^{\lambda_0}$ holds, finding the value of λ to be λ_0 is guaranteed, whereas if T_M holds, finding λ_0 is by no means guaranteed because one might be in a universe where the value of λ is different. So, in terms of subjective probabilities and assuming that we have been able to specify suitable background information B_l – which includes that our universe is life friendly but leaves open the value of λ:

$$P([\lambda_0]|T_M, B_l) < P([\lambda_0]|T_U^{\lambda_0}, B_l), \tag{6.5}$$

with the inequality possibly being substantial.

It is interesting to note that T_M may have comparative advantages over $T_U^{\lambda_0}$ that allow it to partially compensate for the drawback encoded in Eq. (6.5), even before taking into account the fine-tuning considerations F: as discussed in Section 2.2.1, for many constants of fundamental physics whose values are known by observation, physicists have not found any theoretical reason as to why the actual values might be somehow systematically preferred. Notably, in the light of standard criteria of theory choice such as elegance, simplicity, and "naturalness," one would have expected at least some constants to have very different values. The single-universe theory $T_U^{\lambda_0}$ may, therefore, not rank very high in the light of these theoretical virtues.

One can imagine, in contrast, a multiverse theory T_M that is conceptually elegant and has only few, if any, systematically unconstrained constants as input. In that scenario (which, however, is not required for the new fine-tuning argument for the multiverse to go through), it may be plausible to assign a larger prior to T_M than to $T_U^{\lambda_0}$:

$$P(T_M|B_l) > P(T_U^{\lambda_0}|B_l). \tag{6.6}$$

Due to Bayes's theorem, the result of the competition between T_M and $T_U^{\lambda_0}$ is encoded in the ratio:

$$\frac{P(T_M|[\lambda_0], B_l)}{P(T_U^{\lambda_0}|[\lambda_0], B_l)} = \frac{P([\lambda_0]|T_M, B_l)}{P([\lambda_0]|T_U^{\lambda_0}, B_l)} \cdot \frac{P(T_M|B_l)}{P(T_U^{\lambda_0}|B_l)}. \tag{6.7}$$

How the competition between T_M and $T_U^{\lambda_0}$ plays out numerically will, of course, depend on the specific empirical considerations based on which all the probabilities appearing in Eqs. (6.5) and (6.6) are assigned.

Now, using Bayesian language, we can take the impact of the fine-tuning considerations F into account by reconsidering the conditional probability of finding λ_0 *given* T_M in the light of the fine-tuning considerations F and assign it a larger value than while F was still ignored. So, plausibly, if we denote by "$P^F([\lambda_0]|T_M, B_l)$" the revised version of $P([\lambda_0]|T_M, B_l)$, now assigned in the light of the fine-tuning considerations F, we have

$$P^F([\lambda_0]|T_M, B_l) > P([\lambda_0]|T_M, B_l). \tag{6.8}$$

The notation "$P^F([\lambda_0]|T_M, B_l)$" is used instead of "$P([\lambda_0]|T_M, B_l, F)$" because it is doubtful whether taking into account the fine-tuning considerations F can be modeled by Bayesian conditioning. After all, considering F is not a matter of gaining any new evidence but amounts to better understanding the empirical consequences of the theories T_M and T_U – namely, that they are compatible with the existence of life only for very specific configurations of the parameter λ.

In contrast, the conditional probabilities $P^F([\lambda_0]|T_U^{\lambda_0}, B_l)$ and $P([\lambda_0]|T_U^{\lambda_0}, B_l)$ cannot possibly differ from each other as they are both one. As a consequence of Eq. (6.8), the revised version of the inequality Eq. (6.5), taking into account the fine-tuning considerations F, will be less pronounced than the original inequality; i.e., $P^F([\lambda_0]|T_M, B_l)$ will be less tiny compared with $P^F([\lambda_0]|T_U^{\lambda_0}, B_l)$ than $P([\lambda_0]|T_M, B_l)$ was compared with $P([\lambda_0]|T_U^{\lambda_0}, B_l)$.

Now the one key assumption on which the new fine-tuning argument for the multiverse rests is that the fine-tuning considerations F do not have any further, independent, impact on our assessment of the two main rival theories' comparative virtues. Using Bayesian terminology, this key assumption translates into the statement that the ratio of the priors $P^F(T_M|B_l)$ and $P^F(T_U^{\lambda_0}|B_l)$ assigned after taking into account the fine-tuning considerations F will not differ markedly from the ratio of the originally assigned priors $P(T_M|B_l)$ and $P(T_U^{\lambda_0}|B_l)$; i.e.,

$$\frac{P^F(T_M|B_l)}{P^F(T_U^{\lambda_0}|B_l)} \approx \frac{P(T_M|B_l)}{P(T_U^{\lambda_0}|B_l)}. \tag{6.9}$$

Arguably, Eq. (6.9) is a plausible assumption: the fine-tuning considerations F are about the physicochemical preconditions for the existence of life. These are ostensibly unrelated to the systematic virtues and vices of T_M and $T_U^{\lambda_0}$ based on which the assignments of the priors $P(T_M|B_l)$ and $P(T_U^{\lambda_0}|B_l)$ were made. There seems to be no systematic reason as to why the fine-tuning considerations F would inevitably privilege $T_U^{\lambda_0}$ with respect to T_M in any relevant way to be reflected in the priors.

If Eq. (6.9) is assumed, the inequality Eq. (6.8) encodes the only major shift in probability assignments when taking the fine-tuning considerations F into account. Thus, we obtain

$$\frac{P^F(T_M|[\lambda_0], B_l)}{P^F(T_U^{\lambda_0}|[\lambda_0], B_l)} > \frac{P(T_M|[\lambda_0], B_l)}{P(T_U^{\lambda_0}|[\lambda_0], B_l)}; \tag{6.10}$$

i.e., the fine-tuning considerations increase our credence in the multiverse theory T_M. Depending on the result of the original competition between T_M and $T_U^{\lambda_0}$ as encoded in Eq. (6.7) – i.e., dependent on whether we had an initially attractive multiverse theory T_M to begin with – and on how pronounced the inequality Eq. (6.8) is, we may end up with a higher credence in T_M than $T_U^{\lambda_0}$. This completes the new fine-tuning argument for the multiverse.

6.3.3 Attraction and Limits of the New Argument

The main attraction of the new fine-tuning argument for the multiverse is that, unlike the old argument, it is not susceptible to the inverse gambler's fallacy charge: since it treats the life-friendliness of our universe as background knowledge rather than as evidence whose impact we assess, one cannot fault it for failing to consider that the existence of many other universes would not make it more likely that *our* universe is right for life. Another attraction of the new argument in comparison to the old one is that, whereas the old one required assigning prior probabilities $P(T_U|B_0)$ and $P(T_M|B_0)$ from the strange vantage point of someone who is supposedly unaware that there is a life friendly universe, the new one does not do so. The background knowledge B_l can be chosen close to our actual knowledge. It may include all sorts of information about us, our biology, culture, history, and scientific findings. The only bits of information from which it must abstract are those – ideally rather narrowly circumscribed – ones that constrain the parameter configuration λ to its actual value.

The main limitation of the new argument is its reliance on the assumption that the fine-tuning considerations F do not (more than minimally) affect the trade-off between the leading multiverse theory T_M and the leading single-universe theory $T_U^{\lambda_0}$ inasmuch as encoded in the assignment of priors $P(T_M|B_l)$ and $P(T_U^{\lambda_0}|B_l)$. But in the absence of convincing reasons to doubt this assumption, it seems reasonable to suspect that the fine-tuning considerations may indeed make it rational to at least somewhat increase our degree of belief in the hypothesis that we live in a multiverse.

The new fine-tuning argument for the multiverse delivers only incremental and hypothetical support for the multiverse hypothesis. Notably, it requires having independent reasons to take some specific multiverse theory T_M seriously from which we can derive predictions concerning likely values of λ_0. The next part of this book turns to attempts to actually perform such derivations from concrete multiverse theories.

Part III

Testing Multiverse Theories

7

Testing Multiverse Theories
Theoretical Approach

7.1 The Challenge of Testing Multiverse Theories

For practicing physicists, it is ultimately not very interesting whether we live in *some* kind of multiverse in the general sense of the term "multiverse." What really matters is whether some *specific* multiverse theory holds and, if so, whether physicists can find compelling evidence for it. When discussing the fine-tuning argument for a divine designer in Chapter 3, it became clear how important it is to consider specific versions of the designer hypothesis. To realize how much – or, rather, perhaps how little – has been achieved by the fine-tuning argument for the multiverse in its different versions, as discussed in the previous two chapters, it is no less important to study the prospects for empirically assessing specific versions of the multiverse idea – i.e., concrete multiverse theories.

Unfortunately, testing concrete multiverse theories is incredibly difficult, except perhaps for those multiverse theories according to which causal traces of the other universes could be somehow detected in our universe after all. In the landscape multiverse, this, in principle, could be the case in that there might be "bubble collisions" between distinct island universes that would leave detectable imprints on the island universes involved. Identifying those traces and coming to correct conclusions about their origins might be difficult [Aguirre and Johnson, 2011; Salem, 2012]. But there is no reason to doubt that, in principle, it can be accomplished, and if it can be accomplished in practice, obtaining empirical evidence about other universes – inasmuch as colliding "bubbles" deserve to be called individual "universes" at all – is not, in principle, different from obtaining empirical evidence about other remote objects such as, say, distant galaxies or about the very early universe.

But what about multiverse theories according to which there is no causal contact between distinct universes? At least in principle, even such multiverse theories can be tested like other physical theories, namely – as Sean Carroll puts it – by using "abduction, Bayesian inference" [Carroll, 2019, p. 307]. There are simple cases where a procedure of the kind envisaged by Carroll may indeed work rather well. For example, as considered by Greene [2011, chapter 7], one can imagine some multiverse theory T, according to which the value of some parameter (or parameter configuration) λ is very close to a certain specific value λ_0 in *all* subuniverses. In that case, if we find the value of λ to be λ_0 in our own universe and if rival single-universe and multiverse theories do not have this consequence, we can safely regard this finding as supporting the very special multiverse theory T. But this means nothing more than that a multiverse theory may have a good chance of being testable after all if it behaves very *unlike* a typical multiverse theory, namely, in that it predicts some specific value for some specific parameter cosmos-wide.

There is no reason to expect that this – from the point of view of theory assessment – fortunate scenario will be realized for actual candidate multiverse theories like the landscape multiverse. And for theories that entail the existence of universes with very different values of a large number of parameters λ, it is at least prima facie completely unclear how we should assess the evidential impact of our observations with respect to them. In this chapter, I outline the procedure that can be distilled from the extant literature on testing multiverse theories as the one that is widely regarded as the most promising. I will also point out what I think are its most serious problems and risks.

7.2 Toward a Formalization of Anthropic Reasoning

There are two serious roadblocks to empirically testing concrete multiverse theories according to which all, or almost all, parameters are different in the distinct sub-universes of the overall multiverse. The first – "forward-looking" – problem is that any multiverse theory T on its own makes little to no empirical predictions if the multiverse that it entails is sufficiently vast and diverse. Across universes, there may well exist countless observers with qualitatively identical empirical background information to the information we have who are bound to make radically different observations in their futures [Srednicki and Hartle, 2010]. This makes it at least prima facie unclear which future observations such as theory T should be regarded as predicting in light of our present and past.

The second – "backward-looking" – problem is that we may want to account for ("postdict") the values of parameters that we have already measured in our own universe using multiverse theories. But if, according to some multiverse theory T, the values of parameters differ wildly between universes over some wide range, it is again at least prima facie unclear which measured values we should regard as accounted for by T. For example, according to the landscape multiverse scenario, the value of the cosmological constant varies randomly across island universes.

As famously pointed out by Weinberg [1987], only values in some restricted range are compatible with life, and it is unsurprising that the value that we find lies within that range. But the landscape multiverse entails the existence of universes with many or even all values in this "anthropically allowed" range (and many more outside this range), so it seems unclear for which measured values we should regard the landscape multiverse as confirmed and for which ones we should regard it as disconfirmed.

Both problems are especially pressing for types of multiverse theories according to which anything that is allowed to happen by some very general underlying theory will actually happen, perhaps even infinitely often. As Steinhardt points out, this difficulty besets the entire edifice of inflationary cosmology inasmuch as it leads to inflation being eternal:

> [Y]ou should be disturbed. What does it mean to say that inflation makes certain predictions – that, for example, the universe is uniform or has scale-invariant fluctuations – if anything that can happen will happen an infinite number of times? And if the theory does not make testable predictions, how can cosmologists claim that the theory agrees with observations, as they routinely do? [Steinhardt, 2011, p. 42]

What are the strategies that the advocates of multiverse theories have devised to counter this challenge?

7.2.1 Typicality and Self-Locating Indifference

The strategy that is widely considered the most promising for overcoming both problems is to treat a multiverse theory as predicting those observations that *typical* observers make if the theory is true. Following Vilenkin [1995], the term *principle of mediocrity* is widely used for this assumption. The recommendation that we should assume as typical straightforwardly can be seen as a consequence of another assumption, namely, *self-locating indifference*.[1] As expressed by Nick Bostrom [2002], who calls it the *self-sampling assumption* (SSA), it states that one should always reason as if one were randomly sampled from some suitably chosen reference class of observers.

Following Srednicki and Hartle, we may refer to probabilities ascribed to who one might be among the observers in the reference class as *first-person* probabilities and probability distributions ξ that ascribe first-person probabilities as *xerographic distributions*. If there are N observers in the reference class, the uniform xerographic distribution is given by

---

[1] The term *self-identifying indifference* would fit better, but *self-locating indifference* is already widely used, so I stick to this terminology here. Self-locating indifference, as encoded in Eq. (7.1), is defended by Elga [2004] for the very specific case where all observers in the reference class have identical subjective states of affairs.

$$\xi_{ind}(x_i) = \frac{1}{N}, \tag{7.1}$$

where the variable x_i denotes observers in the reference class, and the index "*ind*" stands for "indifference."

Self-locating indifference is equivalent to typicality in the sense that, if one bases one's self-locating credences on the xerographic distribution Eq. (7.1), one will expect to observe what most or "typical" observers in one's reference class will observe. For example, in a multiverse with just two subuniverses A and B, where the parameter λ has the value λ_A in A and λ_B in B, and there are N_A and N_B observers in these universes, respectively, which are all in one's reference class, then self-locating indifference recommends assigning probability

$$\sum_{i_A} \xi_{ind}(x_i) = \frac{N_A}{N_A + N_B} \tag{7.2}$$

to being one of the observers i_A, who populate universe A and find the value of λ to be λ_A. If, for example, $N_A \ll N_B$, then "typical" observers are in universe A, so Eq. (7.2) encodes typicality in that it recommends reasoning as if one could be almost sure to be among the majority of observers – i.e., an observer in universe A. ("Typical" is not a precisely defined notion. Just as the mean and the median of a quantitative variable can differ, what Eq. (7.2) recommends to expect and what observers find that we would intuitively regard as "typical" need not always be the same. Self-locating indifference is a more precisely defined principle, and "typicality" arguably only has appeal in as much as it coincides with it.)

Why would we want to use self-locating indifference? An example by Bostrom illustrates its appeal:

The world consists of a dungeon that has one hundred cells. In each cell there is one prisoner. Ninety of the cells are painted blue on the outside and the other ten are painted red. Each prisoner is asked to guess whether he is in a blue or a red cell. (Everybody knows all this.) You find yourself in one of the cells. What color should you think it is? – Answer: Blue, with 90% probability. [Bostrom, 2002, pp. 59f.]

The answer 90% follows from $\xi_{ind}(x_i) = 1/100$, where $i = 1, \dots, 100$ labels observers by cell number, assuming that the observer reference class should include precisely all prisoners and no one else. As pointed out by Bostrom [2002, pp. 60f.], use of the distribution ξ_{ind} can be motivated in that it is the only credence function that, when simultaneously adopted by all prisoners, prevents them from a sure loss in a well-chosen hypothetical collective bet against them. Srednicki and Hartle [2010] have argued that there is a more compelling motivation available for self-locating indifference – namely, that we should simply *test* typicality as if it were just another scientific theory. I will discuss this proposal in Section 7.2.4 and argue that it fails.

But before that, let us get an impression of extant attempts to extract empirical predictions from concrete multiverse theories. In doing so, we will pay attention to the role that self-locating indifference Eq. (7.1) plays in them.

7.2.2 Successful Examples?

Most extant attempts to derive concrete empirical predictions from some concrete multiverse theory focus on extracting predictions (or, rather, post-dictions) for the value of the cosmological constant from the landscape multiverse scenario of string theory.

In those attempts, the essential conceptual tool is a probability density $f(\Lambda)$ over those values of the cosmological constant Λ that are assumed to be realized in some island universe. The integral over f between any two values Λ_1 and Λ_2 of the cosmological constant yields the probability of finding oneself in an island universe where the value of Λ lies between Λ_1 and Λ_2.

The probability density f can always be decomposed into a probability density $f_V(\Lambda)$ that weighs values of Λ by how many string theory vacua realize them and a weighting factor $W(\Lambda)$. That factor weighs values of Λ in proportion to how many observers on average populate regions where string theory vacua with the respective value of Λ are. What counts as an "observer" will, of course, depend on the chosen reference class. The decomposition reads

$$f(\Lambda) \sim f_V(\Lambda) \cdot W(\Lambda), \tag{7.3}$$

where

$$\frac{W(\Lambda_1)}{W(\Lambda_2)} = \frac{\overline{N}_1}{\overline{N}_2}. \tag{7.4}$$

Here, $\overline{N}_1$ ($\overline{N}_2$) is the average number of observers in vacua that realize the value Λ_1 (Λ_2) of the cosmological constant. Eq. (7.4) evidently implements typicality in analogy with the simple case Eq. (7.2).

String theory is believed to have vacua with both negative and positive values of the cosmological constant. The value $\Lambda = 0$ is not in any way special. Therefore, it is argued, the probability density over vacua should be assumed to be roughly uniform in the close vicinity of $\Lambda = 0$ where observers can exist – i.e., one assumes $f_V(\Lambda \approx 0) \approx const.$

Evaluating the weighting factor $W(\Lambda)$ is trickier. One crude and simple way of doing it is to set that factor to 0 for values of Λ that rule out the existence of observers and to 1 for values of Λ that one assumes to be compatible with the existence of observers. Put differently, one assumes that one's reference class of observers has a fixed number $N = N_0$ elements in vacua with observer-friendly values of Λ and $N = 0$ elements in vacua with observer-hostile values of Λ.

The fine-tuning considerations for Λ pioneered by Weinberg in his seminal article [Weinberg, 1987] indicate an upper boundary of the value of the cosmological constant for compatibility with life around $\Lambda \approx 10^{-120}$ (using natural units), which – according to Martel et al. [1998], gets shifted to somewhat higher values if other parameters are varied. In terms of order of magnitude, the probability density $f(\Lambda)$ will thereby have most of its weight at the order ot 10^{-120} (or even a little higher). The actually measured value of the cosmological constant is closer to 10^{-123} and, as such, disfavored by roughly three orders of magnitude, according to this analysis.

But, of course, this analysis is extremely crude, and there have been attempts to improve it. Their starting point is the observation that values of Λ of the order 10^{-120} would lead to an expansion of the early universe that is much more rapid than the one in our own universe, where the value of Λ is of the order 10^{-123}. In such universes, galaxies will have less time to form and will generally be less massive. Consequently, inasmuch as life requires the existence of big galaxies, it will presumably be less abundant in universes with values of Λ that are as large as 10^{-120}, which means that *typical* observers in the landscape multiverse will presumably inhabit universes with smaller values of Λ. Unfortunately, it is very difficult to make this line of thought quantitatively precise.

The first difficulty is that we have no idea of exactly how many observers there are in the various different subuniverses of the overall multiverse. Attempts to derive predictions for the value of the cosmological constant from the landscape multiverse scenario usually sidestep this difficulty by relying on a *proxy* for observer number such as "proportion of matter clustered in giant galaxies" or "total entropy production." This strategy faces serious problems, however, which will be discussed in Section 8.4.

The second difficulty is that, at least in the landscape multiverse, there seem to be infinitely many island universes, many of which are spatiotemporally infinite [Guth, 2007]. (But see [Ellis and Stoeger, 2009] for dissent.) Accordingly, there are likely infinitely many universes realizing each specific vacuum in the landscape multiverse, and the overall observer number in any observer-admitting universe is also likely infinite. This makes it impossible to straightforwardly compare observer numbers in the different island universes, as required by the typicality principle Eq. (7.2) or by self-locating indifference Eq. (7.1). The difficulty of making observer numbers across universes comparable despite the infinities of the landscape mulitiverse is known as the "measure problem" of cosmology. It is widely regarded as an extremely serious problem, and Tegmark even describes it as the "greatest crisis in physics today" [Tegmark, 2014, p. 314]. The measure problem is discussed in more detail in Section 8.3.

Using a proxy for observer number ("observer proxy" in what follows) and some chosen solution to the measure problem, one can evaluate the weighting factor $W(\Lambda)$ and thereby compute the overall probability distribution over values of the cosmological constant. Calculations using the amount of matter clustered in giant galaxies as observer proxy and the so-called *pocket measure* [Martel et al., 1998] or the *scale-factor measure* [De Simone et al., 2008] result in probability densities $f(\Lambda)$ that peak in the vicinity of the measured value of Λ. The closest agreement between theory and measurement occurs if one assumes that the existence of observers requires very large galaxies. Based on a combined choice of observer proxy and cosmic measure that Bousso et al. [2007] call the "causal entropic principle," these authors derive a result for which the 68% of observers who find non-extreme values of Λ will find values between $4.2 \cdot 10^{-124}$ and $5.8 \cdot 10^{-122}$. The actually measured value, $1.25 \pm 0.25 \cdot 10^{-123}$, agrees very well with this derivation.

Should we regard these computations of probability densities $f(\Lambda)$ that are close to the measured value of Λ as impressive predictive successes of the landscape multiverse? Do they establish that extracting testable predictions from multiverse theories is not only possible but has, in fact, already resulted in empirical confirmations of at least one such theory?

To have a solid basis for answering this question, we must first take a few steps back and revisit the rationale for the application of typicality and self-locating indifference. Notably, we must ask how we could select a reference class of observers for which these assumptions might be plausible.

7.2.3 Problems with Indifference

There are two types of challenges to indifference principles of probability in general, which Weatherson [2005], in a critical response to Elga [2004], applies to indifference principles of self-locating belief like Eq. (7.1). On the one hand, one can criticize indifference principles for relying on the ascription of sharp probabilities to propositions for which there are no good reasons to ascribe to them sharp probabilities at all. The worry, using the terminology of Knight [1921] and Keynes [1921], is that indifference principles treat phenomena that are, in fact, fundamentally *uncertain* as *risky*.

The distinction between risk and uncertainty is not easy to characterize, and the boundary between them is difficult to draw. Knight speaks of risk as something that is "measurable," but it may be more appropriate to characterize it as "quantifiable," in contrast with genuine uncertainty.

There is a widespread view according to which the ascription of probabilities is warranted only in "risky" epistemic situations. For example, Hollands and Wald argue that probabilistic arguments should only be used when the underlying randomness is well understood and a physical mechanism identified:

[P]robabilistic arguments can be used reliably when one completely understands both the nature of the underlying dynamics of the system and the source of its "randomness." Thus, for example, probabilistic arguments are very successful in predicting the (likely) outcomes of a series of coin tosses. Conversely, probabilistic arguments are notoriously unreliable when one does not understand the underlying nature of the system and/or the source of its randomness. [Hollands and Wald, 2002, p. 5]

Philosopher John Norton concurs. As he puts it in a discussion of attempts to extract empirical predictions from scenarios of eternal inflation:

[O]ne cannot assume by default that all uncertainties are to be expressed by probabilities. Rather their expression by probabilities will, in each case, require background conditions that specifically favor it. It is routine for there to be such background conditions. In physical applications these conditions are commonly supplied by the chances of a physical theory. If there is a one in two chance of a head on the toss of a fair coin, or of thermal or quantum fluctuation raising the energy of system, then our uncertainty over whether each happens is well represented by a probability of one half. [Norton, in press, section 5]

Norton [2008] develops a logic of inductive support that he considers applicable when assignments of sharp probabilities are not. In that logic, however, specific predictions (or postdictions) of the sort envisaged by the proponents of multiverse theory can no longer be derived.

Similarly, Weatherson [2005, section 6], in his critical response to how Elga [2004] champions self-locating indifference, proposes an alternative to standard probability theory in self-locating contexts – namely, the theory of *imprecise probabilities*. However, if we follow this recommendation and replace Eq. (7.1) with a statement involving imprecise probabilities, our inferential powers are severely reduced, and the empirical verdicts derived from multiverse theories, as reviewed in the previous section, can no longer be derived.

The second type of challenge brought forward against indifference principles like Eq. (7.1) is that it is often fundamentally unclear with respect to which *reference class* of items, if any, indifference may legitimately be assumed, especially if all candidate reference classes are infinite. A version of this problem was pointed out by Joseph Bertrand [1889] and has become known as "Bertrand's paradox." It has been developed, for example, by van Fraassen [1989] and Shackel [2007].

The second type of challenge is, in fact, closely related to the first. Identifying a preferred reference class of observers is a prerequisite for assigning self-locating probabilities. And once it has been accomplished, the prospects for determining which specific probabilities to assign to its members may seem much brighter, especially if the chosen reference class is finite. One may, therefore, reasonably hope that making progress on the observer reference class problem will also pave the way to the controlled assignment of probabilities to observers.

In fact, cosmologists have long pointed out that typicality and self-locating indifference are ambiguous and that, more generally, the very assignment of self-locating probabilities is problematic as long as there is no clear recipe for selecting the appropriate reference class [Aguirre and Tegmark, 2005; Weinstein, 2006; Hartle and Srednicki, 2007; Azhar, 2014]. The following example, due to Hartle and Srednicki [2007], illustrates this concern (Carroll [2010, p. 225] gives an analogous example that features intelligent lizard beings on a planet orbiting Tau Ceti, and Weinstein [2006, section 2.2] gives an example featuring humans and "aliens").

Imagine that we have an empirically well-motivated theory T_J that entails that there are hitherto unknown intelligent observers in the atmosphere of Jupiter ("Jovians"). In fact, there are many more Jovians than there are humans on Earth; i.e., $N_J \ll N_H$ for their numbers. As an alternative to T_J, we consider a "null hypothesis" T_0, according to which there are no Jovians but only the N_H humans. (It is assumed that there are no other candidate observers besides humans and Jovians that we need to consider.)

The problem of observer reference class choice becomes manifest in the question of whether we should include the Jovians in the reference class. The answer to this question may dramatically impact our preference between T_J and T_0. Assume, for the sake of simplicity, that there are no observers outside our solar system, and suppose that we do include the Jovians in the reference class. Then, if T_J is true, we are highly *atypical* observers in the reference class because most reference class members are Jovians rather than humans. In contrast, if T_0 is true, Jovians do not exist, and the reference class contains only humans, which makes us typical. Typicality suggests penalizing theories according to which we are atypical, so, by the assumptions made, it suggests discarding T_J even if our empirical evidence otherwise happens to be neutral between T_J and T_0 or even slightly favors T_J.

The unattractive preference for T_0 disappears if we do not include the Jovians in the reference class. But, assuming that the hypothesized Jovians are sentient and intelligent, it is unclear on what basis one could justify their exclusion. A well-motivated demarcation criterion for observer reference class membership is clearly needed.

As pointed out by Garriga and Vilenkin [2008], it is possible to avoid the unattractive systematic preference for T_0 over T_J without excluding the Jovians from the reference class by invoking the controversial *self-indication assumption* (SIA) [Bostrom, 2002]. According to the SIA (to be discussed in more detail in Section 9.2.1), which assumes a Bayesian framework of theory assessment, we should assign *prior* probabilities in such a way that theories are privileged and/or disfavored in proportion to the number of observers whose existence they entail. When applied to the humans and Jovians scenario, the SIA precisely cancels out the intuitively implausible systematic preference for T_0 over T_J, and it is advocated independently by Olum [2002]. However, as acknowledged by Garriga and Vilenkin, the SIA has highly counterintuitive consequences in other domains, notably the so-called *presumptuous philosopher problem* [Bostrom, 2002, p. 124], which will be discussed in detail in Section 9.2.3. Accepting the SIA would, in any case, not solve the demarcation problem for observer reference class membership.

Cosmologists Srednicki and Hartle [2010] propose that we could solve the observer reference class problem by empirically *testing* typicality with respect to various candidate reference classes and, in the end, opt for the competitively most successful reference class. The following section reviews Srednicki and Hartle's approach; the subsequent section criticizes it.

7.2.4 Testing Typicality?

Srednicki and Hartle stage their proposal to test typicality with respect to different reference classes by combining physical theories T and xerographic distributions ξ in dyads (T,ξ) that they call "frameworks." As they argue, testing such frameworks can be conveniently modeled along Bayesian lines. The starting point is a prior "third-person" probability distribution $P(T,\xi)$ over candidate frameworks (T,ξ). Using Bayes's theorem, first-person probabilities $P^{1p}(T,\xi|D_0)$, conditional with respect to our complete background information D_0, can be obtained from the third-person probabilities using Bayes's theorem (see eq. (4.1) in [Srednicki and Hartle, 2010]):

$$P^{1p}(T,\xi|D_0) = \frac{P^{1p}(D_0|T,\xi)P(T,\xi)}{\sum_{(T,\xi)} P^{1p}(D_0|T,\xi)P(T,\xi)}. \tag{7.5}$$

A further crucial assumption by Srednicki and Hartle is that we can treat the fundamental physical theory T and the xerographic distribution ξ as independent in that the third-person probability distribution over frameworks (T,ξ) factorizes into a contribution P_{th} over theories and a contribution P_{xd} over xerographic distributions:

$$P(T,\xi) = P_{xd}(\xi) \cdot P_{th}(T), \tag{7.6}$$

where, plausibly,

$$P_{xd}(\xi) = \sum_j P(T_j,\xi), \tag{7.7}$$

and

$$P_{th}(T) = \sum_j P(T, \xi_j), \qquad (7.8)$$

By Eq. (7.6), we can assess the impact of incoming empirical evidence with respect to P_{xd} and P_{th} separately. Notably, as emphasized by Srednicki and Hartle, this has the consequence that we can "compete" different xerographic distributions against each other for the same theory T. Since typicality with respect to any specific reference classes can be expressed as indifference $\xi_{ind}(x_i) = 1/N$ over the observer reference class, this "competition" promises an empirical solution to the observer reference class problem with the victorious reference class selected as the one to be used in future empirical tests of multiverse (and other cosmological) theories.

The xerographic distribution $\xi(x_i)$ is itself a probability distribution, so the probabilities ascribed to xerographic distributions by $P_{xd}(\xi)$ are second-order. There is no difficulty in principle with second-order probabilities. For example, hypotheses H_θ concerning the unknown bias θ of some (potentially biased) coin can be individuated in terms of their probability ascriptions to the various possible sequences of outcomes. Accordingly, probabilities $P(H_\theta)$ defined over such hypotheses concerning coin bias are second order. They are obviously well defined and can be tested by repeated coin tossing and evaluating the toss outcomes.

However, there is an important difference between hypotheses H_θ concerning coin bias and xerographic distributions ξ: while there supposedly exists some actual – perhaps unknown – bias θ of the coin (defined in terms of long-term outcome frequencies) so that precisely one H_θ is true, there is no "true" xerographic distribution ξ, except in the uninteresting and trivial sense in which each observer x_j has this role played by their characteristic function $\chi_j(x_i) = \delta_{i,j}$.

This fundamental difference between hypotheses concerning unknown coin biases and xerographic distributions becomes relevant when it comes to updating procedures with respect to ξ and H_θ, respectively. Whereas the evidential impact of data concerning coin toss outcomes on probability assignments over the hypotheses H_θ can be conveniently modeled by Bayesian updating of $P(H_\theta)$, it turns out that the impact of self-locating information is adequately modeled only by updating the xerographic distribution ξ *itself* rather than by updating any probability distribution $P_{xd}(\xi)$ over xerographic distributions.

To see this, consider again Bostrom's dungeon, which hosts 90 prisoners in blue cells and 10 prisoners in red cells. Suppose that a prisoner somehow finds out that she is in a blue cell. According to the approach suggested by Srednicki and Hartle, her rational posterior first-person probability $P^{>,1p}(\xi)$ after finding this out, which is given by her conditional first-person prior probability $P^{1p}(\xi|\text{My cell is blue})$, evaluated at fixed theory T for which $P_{th}(T) = 1$, is obtained from Eq. (7.5) as

$$P^{1p}(\xi|\text{My cell is blue}) = \frac{P^{1p}(\text{My cell is blue}|\xi)P_{xd}(\xi)}{P^{1p}(\text{My cell is blue})}. \quad (7.9)$$

Now suppose that the prior $P_{xd}(\xi)$ over xerographic distributions ascribes a non-zero probability to at least one xerographic distribution ξ_0 that has nonvanishing support in both blue and red cells. For example, suppose that the prior $P_{xd}(\xi_{ind})$ assigned to $\xi_{ind}(x_i) = 1/100$ is non-zero (possibly 1).

By assumption, ξ_0 ascribes non-zero probability to being in a blue cell; i.e., $P^{1p}(\text{My cell is blue}|\xi_0)$ is non-zero. By Eq. (7.9), $P^{1p}(\xi_0|\text{My cell is blue})$ must be non-zero as well. This is problematic, however: by assumption, ξ_0 ascribes non-zero probability to being an observer in a red cell, and this conflicts with one's determinate knowledge that one's cell color is blue. Updating in accordance with Eq. (7.9) thus leads to an awkward situation where one is, on the one hand, completely certain that one is in a blue cell while simultaneously entertaining, at least in some sense, the possibility that one is in a red cell.[2]

The problem is not confined to simple examples like Bostrom's dungeon but also appears in applications of Srednicki and Hartle's framework to actual cosmological problems. For example, assume that we want to account for the observed value of the cosmological constant Λ and assign a prior $P_{xd}(\xi)$ that is non-zero (perhaps even 1) to the uniform xerographic distribution ξ_{ind} over observers who witness the full anthropically allowed range Δ_Λ across universes. Having measured Λ in our own universe and having found it to be within some finite proper subrange $\delta_\Lambda \subset \Delta_\Lambda$, the posterior first-person probability $P^{>,1p}(\xi_{ind}) = P^{1p}(\xi_{ind}|\Lambda \in \delta_\Lambda)$ assigned to ξ_{ind} is non-zero by Bayes's theorem because ξ_{ind} is non-zero over the members of δ_Λ. Assigning a non-zero posterior to ξ_{ind} in that situation is incoherent, however, because ξ_{ind} ascribes non-zero probability to Λ lying in Δ_λ but *outside* δ_Λ, contrary to measurements of Λ according to which $\Lambda \in \delta_\Lambda$.

The adequate procedure for taking into account the impact of self-locating information like "My cell is blue" or "I'm in a universe with $\Lambda \in \delta_\Lambda$" does not seem to be Bayesian updating of any second-order probability distribution $P_{xd}(\xi)$. A much simpler and better option is to directly update some prior xerographic

---

[2] No analogous difficulty arises with respect to hypotheses H_θ concerning coin toss bias. Here, the posterior $P^>(H_\theta) = P(H_\theta|D)$, where D are outcome data concerning observed tosses, is given by

$$P(H_\theta|D) = \frac{P(D|H_\theta)P(H_\theta)}{P(D)}.$$

This is unproblematic because being certain about D is not in tension with assigning a non-zero (posterior) probability to H_θ. Whether H_θ obtains concerns the limiting long-run frequency behavior of the coin, assumed to be a determinate matter, which is not in any possible conflict with the outcome data D.

distribution ξ_{prior} *itself*. (If there are various candidate priors ξ_k between which it is difficult to decide, one can use a weighted average $\xi = \sum_k w_k \xi_k$, where the w_k can be chosen such as to correspond to the $P_{xd}(\xi_k)$ in Srednicki and Hartle's scheme.)

This option immediately delivers the intuitively correct and plausible verdict for the applications discussed. For example, if in Bostrom's dungeon a prisoner starts with the uniform prior $\xi_{ind}(x_i) = 1/100$ over all prisoners, ordinary Bayesian conditioning with respect to "My cell is blue" yields the attractive posterior $\xi^>(x_i) = \xi_{ind}(x_i|\text{My cell is blue})$ given by the conditional prior

$$\xi_{ind}(x_i|\text{My cell is blue}) = \frac{\xi_{ind}(\text{My cell is blue}|x_i) \cdot \xi_{ind}(x_i)}{\xi_{ind}(\text{My cell is blue})}$$

$$= \begin{cases} \frac{1 \cdot 1/100}{9/10} & \text{if } x_i \text{ is a blue} - \text{cell observer} \\ \frac{0 \cdot 1/100}{9/10} & \text{otherwise} \end{cases}$$

$$= \begin{cases} 1/90 \text{ if } x_i \text{ is a blue} - \text{cell observer} \\ 0 \text{ otherwise} \end{cases} \qquad (7.10)$$

In this updating scheme, unlike in the one that derives from the framework proposed by Srednicki and Hartle, typicality as encoded in ξ_{ind} is not empirically "tested" in that it is not treated as a hypothesis in analogy to some theory T that one treats as confirmed or disconfirmed by evidence. Prima facie, it seems to be used as a primitive starting point for which no further justification is given. Having abandoned the idea of solving the observer reference class problem by testing typicality with respect to candidate reference classes, we will see in what follows how paying attention to one's background information D_0 allows one – at least in principle – to choose the reference class in a non-arbitrary way.

7.3 The Background Information Constraint

If there are observers with psychological states that are subjectively identical to one's own and one knows this, then, Elga [2004] argues, self-location indifference with respect to the reference class that includes precisely those observers is a reasonable assumption. In a similar spirit, Garriga and Vilenkin suggest to include in the observer reference class precisely those observers with "identical information content" as oneself [Garriga and Vilenkin, 2008, abstract].

Unlike Elga, Garriga and Vilenkin never commit to an explicitly psychological account of what constitues "information content" D_0. In fact, the results of applying Garriga and Vilenkin's recommendations, as discussed in the following paragraphs, may depend on whether one adheres to an "internalist" conception of D_0, along the lines of Elga's principle, or favors a more "externalist" conception, according to which D_0 is information that is in some, perhaps idealized, sense physically present in or for our epistemic agent, or at least physically accessible for her. Since we will encounter significant challenges to implementing the "information content"-based approach to multiverse theory testing explored in what follows that are independent of how the notion is understood, I will therefore leave it open along which lines exactly one should think of D_0.

Using the notion of "information content" D_0, the central principle for reference class choice according to Garriga and Vilenkin [2008] is the following: given an observer's full (first- *and* third-person) "information content" D_0, the appropriate reference class to use is the one that contains precisely those whose background information is D_0.

This suggestion is helpful in some circumstances but delivers overly restrictive verdicts in others. Consider again Bostrom's dungeon scenario: one would expect that prisoners in different cells have different memories and/or current states of knowledge, but this will not, per se, make it irrational for them to assign the uniform distribution $\xi(x_i) = 1/100$ over prisoners. What *would* make it irrational is if their background information D_0 allowed them to rule out being some of the 100 prisoners. This simple idea is encoded in the following *background information constraint* (BIC) on the observer reference class, which I think is the appropriate principle guiding observer reference class choice in anthropic reasoning:

(BIC) Given background information D_0, include in the observer reference class precisely those observers who you possibly *could be* in view of D_0.

Typicality is only attractive if one's background information D_0 is judged to be evidentially *neutral* with respect to who one might be among those who one possibly *could be*, given D_0. If this neutrality constraint is not fulfilled, it is obviously more attractive to use a nonuniform xerographic distribution instead of the uniform one, though still one based on the background information constraint BIC. In what follows, I will assume for the sake of simplicity that the neutrality constraint is fulfilled.

Let us assume that it can, in practice, be determined who among all observers that exist according to some theory T one could possibly be in view of one's background information D_0. The resulting observer reference class to be used according to the BIC will then, in general, depend strongly on the theory T. Different fundamental theories T_1 and T_2 will typically differ on which and how many observer-type physical structures exist and on how many of them one could possibly be, given that D_0 is one's background information. Accordingly, xerographic distributions associated with T_1 and T_2 will, in general, differ as well. We must therefore assign an index for the theory T to the xerographic distribution so that it becomes $\xi_T(x_i|D_0)$.

By the definition of conditional probability, the first-person probability $P^{1p}(T, x_i|D_0)$ can be written in product form as

$$P^{1p}(T, x_i|D_0) = P^{1p}(x_i|T, D_0) \cdot P^{1p}(T|D_0), \tag{7.11}$$

which replaces Srednicki and Hartle's factorization assumption Eq. (7.6).

The first factor on the right-hand side of Eq. (7.11) specifies the probability of being x_i, assuming T's truth and one's background information D_0. This makes it plainly equivalent with the xerographic distribution $\xi_T(x_i|D_0)$, which also specifies the probability of being x_i, given D_0, if T holds. The second factor on the right-hand side of Eq. (7.11) applies only to the "third-person" fact of which theory T holds. One can identify it with the third-person probability distribution $P_{th}(T|D_0)$ over theories. In virtue of these identities, Eq. (7.11) becomes

$$P^{1p}(T, x_i|D_0) = \xi_T(x_i|D_0) \cdot P_{th}(T|D_0). \tag{7.12}$$

The following section illustrates how BIC, using Eq. (7.12), can, in principle, solve the observer reference class problem by applying it to the humans and Jovians scenario.

7.4 The Reference Class Problem (Formally) Resolved

Let us recapitulate why we should expect that there are likely no Jovians if we include the Jovians in our observer reference class. The indifference-expressing xerographic distribution in this case is $\xi_{T_J, ind}(x_i|D_0) = 1/(N_J + N_H)$ for T_J and $\xi_{T_0, ind}(x_i|D_0) = 1/N_H$ for T_0, where N_J is the number of Jovians according to T_J and N_H is the (known) number of humans according to both T_J and T_0.

From Eq. (7.12), we obtain the ratio of the first-person probabilities $P^{1p}(T_J, \text{I am human}|D_0)$ and $P^{1p}(T_0, \text{I am human}|D_0)$ as

$$\frac{P^{1p}(T_J, \text{I am human}|D_0)}{P^{1p}(T_0, \text{I am human}|D_0)} = \frac{N_H}{N_J + N_H} \cdot \frac{P_{th}(T_J|D_0)}{P_{th}(T_0|D_0)}. \tag{7.13}$$

By assumption, $N_H/(N_J + N_H) \ll 1$ because $N_H \ll N_J$. As a consequence, unless the third-person probabilities $P_{th}(T_J|D_0)$ and $P_{th}(T_0|D_0)$ happen to strongly favor T_J over T_0, the first-person probabilities $P^{1p}(T_J|D_0, \text{I am human})$ and $P^{1p}(T_0|D_0, \text{I am human})$ will strongly favor T_0 over T_J. The observation that we are humans thus supports the null hypothesis T_0 according to which there are no Jovians over the alternative hypothesis T_J according to which there are Jovians.

But should we include the Jovians in the observer reference class according to BIC? To answer this question, we need to determine whether we possibly could be Jovians, given our background information D_0. Now, our complete *actual* background information D_0, which includes all (scientific and everyday) data that we currently happen to have, plainly includes the fact that we are humans and live on Earth. Plainly, it not only rules out that we are Jovians but even specifies precisely who we are among humans – identifying each of us in terms of name, date of birth, place of birth, etc. Our actual background information thus effectively narrows down who we are to some specific human observer x_j. So, using our actual background information as D_0, we obtain from Eq. (7.12)

$$\frac{P^{1p}(T_J, \text{I am human}|D_0)}{P^{1p}(T_0, \text{I am human}|D_0)} = \frac{P_{th}(T_J|D_0)}{P_{th}(T_0|D_0)}, \tag{7.14}$$

which yields no preference in first-person probabilities for T_0 over T_J when taking into account one's being human.

But perhaps we should consider operating on the basis of significantly more restricted background information D_0 that does *not* give away the fact that we are human. Indeed, to assess the impact of certain data – in this case, of the fact that we are human – we must do so on the basis of background information D_0 that does not include them. And if some hypothetical amount of background information D_0 has been obtained by abstracting from our being human, it may well be compatible with its bearer being Jovian. Then BIC would dictate using a reference class that indeed does have Jovians as members. For such background information D_0, the argument centered around Eq. (7.13) can be run, and the systematic preference for T_0 can apparently be vindicated.

However, while abstracting from bits of our actual background information to arrive at a suitable D_0 is, in principle, possible, the actual, practical outcome of this abstraction process quickly becomes dubious and unclear. We may ask ourselves, for example, whether hypothetical background information D_0 that is compatible with us being Jovians should plausibly include data gathered at particle accelerators built and operated by humans and, if some such data should be included, which data to include. The answer to this question has the potential to very strongly impact the appropriate choice of observer reference class.

Moreover, since any candidate background information D_0 arrived at by abstracting from the fact that we are human is so radically impoverished with respect to our *actual* background information, it is difficult to estimate which third-person probability assignments $P_{th}(T_J|D_0)$ would be rational to assign on its basis. Our ordinary plausibility verdicts are probably no reliable guide if the hypothesized background information D_0 differs so massively from our actual state of knowledge. But unless we have a prima facie reasonable case for assigning values

$P_{th}(T_J|D_0)$ and $P_{th}(T_0|D_0)$ of specific orders of magnitude, we have, in light of Eq. (7.12), no strong case for first-person probabilities $P^{1p}(T_J|B, \text{I am human})$ and $P^{1p}(T_0|B, \text{I am human})$ that strongly favor T_0 over T_J. To conclude, BIC delivers a formal solution to the observer reference class problem that, when applied to the humans and Jovians scenario, does not yield the unattractive verdict that T_0 is to be systematically preferred over T_J. It does not resolve the practical difficulties of determining the appropriate reference classes in concrete cases.

8

Testing Multiverse Theories
Approaching Practice

8.1 Implementing the Background Information Constraint

In order to actually derive empirical predictions from specific multiverse theories using the BIC, we must reflect on which background information D_0 we should use when trying to account for the measured values of fundamental parameters in terms of those theories. The most obvious choice may seem to be our full actual background information and, correspondingly, the reference class of all observers who we could possibly be, given our actual background information and a complete description of the world, assuming that the theory T that we consider is correct.

However, if we want to account for the value of some already measured parameter, this is not an option because it would prevent us from treating that value as a nontrivial prediction of T. For example, if we want to account for the measured value of the cosmological constant using the landscape multiverse scenario, we must use a reference class of observers who find different values for that constant and then determine whether the value that we find is a typical one among those found by observers in that reference class.

There is probably no single best recipe for singling out any specific proper part D_0 of our actual background information as ideally suited for this task. The following two desiderata, however, seem reasonable to request and may serve as rough guidelines:

- First, it should be practically feasible to determine unambiguously which and how many observers one could possibly be, given the chosen background information D_0, for any theory T that one wishes to test.
- Second, it should be practically possible to reasonably motivate one's assigned third-person probabilities $P_{th}(T|D_0)$ for the chosen D_0 and any theory T that one wishes to test.

Satisfying any of the two desiderata for some realistic candidate multiverse theory T like the landscape multiverse scenario and some chosen background information D_0 is an extremely big challenge.

Let us suppose that we want to test some multiverse theory T with respect to whether it can account for the value of some parameter λ as measured in our own universe. Let us further suppose that we have successfully disentangled any knowledge we may happen to have about the value of λ from the rest of our physical knowledge D_0. We can then use this D_0 as background information based on which we assess different theories with respect to their ability to predict the value of λ. The BIC then requires us to determine the properties of all universes that exist according to T and harbor observers who we could possibly be, given our assumed background knowledge D_0.

But this means, in effect, that we do not only have to determine the higher-level physical and chemical laws of all universes with different values of λ that we could be in, given D_0, but also the turns biological evolution is likely to take in those universes.

Our chances to keep this task manageable are probably best if we assume background knowledge D_0 that is as close as possible to our full actual background knowledge. Ideally, we would therefore use background knowledge D_0 that differs from our actual background knowledge only inasmuch as it does not include the value of the parameter λ that the theory T under consideration is supposed to account for. For example, if we try to account for the value of the cosmological constant Λ, then our prospects for correctly identifying the observer reference class we should use according to the BIC are greatest if we choose D_0 such that it leaves open the value of Λ but fixes the values of as many other parameters as possible to those in our universe. Since, for many orders of magnitude of Λ, its specific value has no great effect on nuclear and atomic physics or on chemistry, this means that we need to consider only universes with the same nuclear and atomic physics and the same chemistry as our own universe (though the abundance of nuclei may depend strongly on Λ because its value has an impact on the duration of cosmic eras when nuclei form).

Our task does not thereby become easy, though, because we must still determine which values of Λ will give rise to observers who we could possibly be, given D_0 as our background evidence – i.e., which universes have galaxies sufficiently like the Milky Way and stars sufficiently like the sun, orbited by planets sufficiently similar to Earth, populated by creatures sufficiently like us. We can expect this to be difficult because we do not know how abundant humanlike observers in Milky Way-like galaxies are even in our universe, let alone how abundant they are in universes where galaxy formation proceeds differently because Λ is different.

However, one crucial drawback of choosing background knowledge D_0 that is very close to our actual background knowledge is that it may trivially narrow down the reference class of observers to observers who live in universes where the value of the parameter λ to be accounted for is extremely close to the value in our own universe. For example, if – in trying to account for the value of the cosmological constant Λ – we use background knowledge D_0 that contains detailed information about the sizes of galaxies in our universe, it may thereby "give away" so much information about galaxy formation that we can essentially determine the value that Λ must have. But if D_0 is so specific that it determines the value of Λ independently of the theory T, there is little we can learn from the attempt to account for Λ using T in that instance.

In other words, if – for some suggested background knowledge D_0 – the probability density $f(\Lambda)$ is strongly peaked close to the value that Λ has in our universe for *any* theory T that we may consider, then using that D_0 is not very helpful for competing those theories against each other. We will have to abstract from more aspects of our actual knowledge and work with more impoverished background knowledge D_0 for which candidate theories T might appreciably differ as regards the value of Λ to be expected if we want to discriminate between those theories T.

Unfortunately, for background knowledge D_0 that differs substantially from our actual background knowledge by abstracting from much that we actually know, the BIC requires us to consider universes that have very different effective (higher-level) physical laws than our own. Those universes differ to some degree from our own with respect to nuclear physics, atomic physics, and chemistry. Estimating what complex structures like biological organisms will be like in those universes – and how abundant they will be – is not only challenging but often downright impossible. Are there prospects for overcoming this challenge?

8.2 Observer Proxies

The methods that are used in cosmological practice to derive predictions from multiverse theories despite these difficulties can be seen as conforming to the BIC, at least in spirit. To make this transparent, it is helpful to decompose the probability distribution $f(\lambda)$ somewhat differently than in Eq. (7.3), where this was done for the cosmological constant specifically – namely, as

$$f(\lambda) = \xi_T(\lambda|D_0) \sim \bar{n}_{O_M}(\lambda|D_0) \cdot n_M(\lambda). \tag{8.1}$$

Here, $\xi_T(\lambda|D_0)$ is a shorthand for $\sum_{x_i} \xi_T(x_i|D_0)$, where the sum is taken over all observers x_i who are located in regions where the specific value λ is realized.

The index M labels regions (e.g., subuniverses) of the overall multiverse, the number $n_M(\lambda)$ counts regions M with the parameter value λ, and $\bar{n}_{O_M}(\lambda|D_0)$ is the mean number of observers who one could be, given that D_0 is one's background knowledge, and who are located in regions M where the parameter value is λ.

As we have just seen, making a well-informed estimate about how many reference class observers associated with D_0 are there in regions M with λ is incredibly hard. In practice, as explained in Section 7.2.2, observer proxies are employed to nevertheless give estimates of $\bar{n}_{O_M}(\lambda|D_0)$ that, it is hoped, are at least of the right order of magnitude.

The observer proxy that has been most widely used in attempts to account for the measured value of the cosmological constant is *proportion of baryon matter clustered in giant galaxies* [Martel et al., 1998]. The motivation for using this proxy comes from considering universes with the same parameters as ours except for the fact that they have a 100–1,000 times larger cosmological constant Λ. In such universes, the larger Λ creates a more accelerated early cosmic expansion, which leaves less time for galaxy formation than there was in our own universe. Galaxies in such universes will in general be smaller than those in our own universe, and their comparatively small masses may not suffice to create the supernovae explosions necessary for the formation of heavier elements that are needed for the existence of observers similar to us. One may therefore expect universes with larger Λ to be less densely populated by observers than our own universe – whose observer density, by the way, we do not know – and this leads to the idea of treating the proportion of baryon matter clustered in giant galaxies as a proxy for how many observers there are.

However, as pointed out by Bousso et al. [2007], this proxy has several problems: first, it is not clear why baryons as opposed to, say, photons or some specific type of nuclei on which our existence also depends should be regarded as especially relevant; second, even if we decide that baryons are of overarching importance, it is not clear why precisely the *proportion* of baryons that cluster into big galaxies should matter so much more than, say, the absolute number of baryons clustered in big galaxies; third, it is not clear whether one can define "baryons" at all in universes that realize generic string theory vacua – notably, string theory vacua with considerably different effective laws. In those universes, *proportion of baryon matter clustered in giant galaxies* may not even be well defined, let alone be useful as an observer proxy.

As an alternative observer proxy that does not suffer from these problems, Bousso et al. [2007] propose *amount of entropy production* in the so-called *causal diamond*. Entropy production is well defined independently of the effective physical laws of a given universe, and, by being related to the free energy, is suspected by Bousso et al. to be a good measure of intuitive *complexity*. Conversely, where entropy production is zero, thermal equilibrium obtains, and there are no observers.

Unfortunately, this observer proxy has its weaknesses as well. Most importantly, while it seems reasonable to assume that observers can only exist in a space-time region with entropy production, there is no reason to believe that entropy production reliably correlates – let alone correlates *linearly* – with observer number. Notably, it seems entirely possible that there are regions of space-time with considerable entropy production and, hence, much "complex structure" but without any observers that belong to the reference class associated with one's chosen background knowledge D_0.

The Humans and Jovians scenario illustrates this worry – which, in fact, applies to any observer proxy of the sort just discussed. In that scenario, which candidate cosmological theory T_0 or T_J one regards as better supported by the evidence may strongly depend on whether a given group of candidate observers – in this case, the Jovians – is included in the observer reference class or not. How this type of question can, in principle, be settled by appeal to the BIC has been shown in the previous section. However, observer proxies like amount of matter clustered in giant galaxies or entropy production are by far too crude as measures to take into account such vital information as to whether the hypothetical Jovians are to be included in the reference class of observers or not.

If, in assessing our priors $P(T|D_0)$, we use background knowledge that, for example, includes the information that we are *not* Jovians, then this information must be somehow reflected in the observer proxy that we use – i.e., that proxy must tack the number of humans, but not that of Jovians. The physical quantities used as proxies are unable to reflect such information and, hence, are a far cry from fulfilling this reasonable desideratum.

Just how dramatically the predictions derived from the landscape multiverse depend on the observer proxy being used is strikingly demonstrated by Starkman and Trotta [2006]. For a proxy that they dub *maximum allowed number of observations*, they show that values of the cosmological constant Λ, which are dramatically *smaller* than the one measured in our universe, are actually favored. As they acknowledge, this proxy is no more compelling than any of the other proxies suggested in the literature, but, as they convincingly argue, it is also not dramatically *less* compelling.

The right way to argue for the use of one specific candidate observer proxy over others would be to highlight why that proxy, unlike others, can be expected to reliably correlate (ideally linearly) with the number of observer in the reference class picked out by the assumed background knowledge D_0 in the light of the BIC. But, unfortunately, choosing observer proxies in this way does not seem feasible in practice. For predictions of Λ, this means that, due to their strong dependency on the chosen observer proxy, they are unfortunately not reliable.

8.3 Cosmology's Measure Problem

Having discussed the difficulties of estimating the average number $\bar{n}_{O_M}(\lambda|D_0)$ in some specified type of region M, let us turn to the difficulty of selecting an appropriate type M. For the landscape multiverse, island universes may seem to be the most natural candidate regions M, but this idea runs into difficulties: according to the consensus view of the landscape multiverse, there are infinitely many island universes, many of them with an infinite spatiotemporal extension. Observer-hospitable island universes will, thus, in general, be populated by infinitely many observers. This is a problem because ratios between infinite numbers (of the same cardinality) are not unambiguously defined, at least not without specifying a *regularization procedure*: a prescription for comparing finite numbers first and only then taking the limit to infinity in a controlled way, so that ratios remain finite.

Proponents of eternal inflation take this problem very seriously. Alan Guth, the pioneer of inflationary cosmology, illustrates it as follows:

To understand the nature of the problem, it is useful to think about the integers as a model system with an infinite number of entities. We can ask, for example, what fraction of the integers are odd. Most people would presumably say that the answer is 1/2, since the integers alternate between odd and even. That is, if the string of integers is truncated after the Nth, then the fraction of odd integers in the string is exactly 1/2 if N is even, and is $(N+1)/2N$ if N is odd. In any case, the fraction approaches 1/2 as N approaches infinity. However, the ambiguity of the answer can be seen if one imagines other orderings for the integers. One could, if one wished, order the integers as

$$1, 3, 2, 5, 7, 4, 9, 11, 6, \ldots, \tag{8.2}$$

always writing two odd integers followed by one even integer. This series includes each integer exactly once, just like the usual sequence $1, 2, 3, 4, \ldots$). The integers are just arranged in an unusual order. However, if we truncate the sequence shown in Eq. (8.2) after the Nth entry, and then take the limit $N\infty$, we would conclude that 2/3 of the integers are odd. Thus, we find that the definition of probability on an infinite set requires some method of truncation, and that the answer can depend nontrivially on the method that is used. [Guth, 2007, section 4]

Similar analogies to illustrate the problem of handling the infities that arise in eternal inflation are given by Tegmark [2014, p. 16], Vilenkin [2007, p. 6779], and Steinhardt [2011, p. 42].

For the natural numbers, there may seem to be a preferred and "natural" method of truncation, dictated by their conventional ordering. It delivers the intuitively correct share $1/2$ for the odd numbers. For the eternally inflating "multiverse" of island universes, in contrast, it is far less clear which, if any, is the preferred truncation procedure for space-time. As Guth explains:

In the case of eternally inflating spacetimes, the natural choice of truncation might be to order the pocket universes in the sequence in which they form. However, we must remember that each pocket universe fills its own future light cone, so no pocket universe forms in the future light cone of another. Any two pocket universes are spacelike separated from each other, so some observers will see one as forming First, while other observers will see the opposite. One can arbitrarily choose equal-time surfaces that foliate the spacetime, and then truncate at some value of t, but this recipe is not unique. In practice, different ways of choosing equal-time surfaces give different results. [Guth, 2007, section 4]

There are two types of reactions to the problem of infinities in eternal inflation, the "measure problem": one is to give up on trying to make any probabilistic predictions (or postdictions) from the landscape multiverse (or other multiverse proposals that suffer from similar problems); the other is to try to determine a systematically preferred truncation procedure and treat the theory as predicting what this truncation procedure delivers together with an appropriately chosen observer proxy. I discuss these options in turn.

8.3.1 Giving Up on Probabilistic Predictions?

One possible reaction to the measure problem is that it makes any scenario involving eternal inflation effectively untestable. This view underlies the general attack on inflationary cosmology by Ijjas, Loeb, and Steinhardt, as mentioned in Section 1.2.3. How devastating the measure problem in eternal inflation actually is for inflationary cosmology in general depends, of course, on whether there are compelling models of inflation according to which it is non-eternal. As discussed in Section 1.2.1, Mukhanov [2015], for example, provides an example of such a model. According to Martin [in press, section 7C], that model is in agreement with the [Planck Collaboration, 2016] findings on the CMB fluctuations. Martin goes on to provide further arguments [Martin, in press, section 7C] for not regarding the systematic difficulties encountered by testing eternal inflation as a reason to reject inflationary cosmology as a whole.

Nevertheless, as far as eternal inflation as a multiverse-generating mechanism is concerned, the measure problem is troubling in any case. Norton [in press] argues that it makes the application of ordinary probability theory impossible and permits only what calls the "infinite lottery inductive logic": the rational approach to a (hypothetical) infinite lottery. The infinite lottery inductive logic does not make predictions impossible in principle, but it "does affirm that predictions useful to deciding for or against eternal inflation are precluded" [Norton, in press, abstract].

My own view is that this stance is, in the end, the right one to take: the measure problem does seem to make scenarios that involve eternal inflation untestable, especially when the challenge of selecting an appropriate observer proxy is also taken into account. I elaborate on this view in Section 8.3.3. First, let us consider the specific regularization proposals that have been made in order to solve the measure problem.

8.3.2 Selecting a Specific Measure: The Options

In some cases, it may seem obvious to which results acceptable regularization procedures would lead. The "natural" ordering of the natural number sequence, as we saw earlier, provides an example. Another example might be an infinite regular chessboard with alternating white and black squares. Acceptable truncation procedures would presumably deliver the verdict that the numbers of white and black squares are equal. One that indeed produces this result is to first consider a finite – say, 68 times 68 squares – chessboard and then successively add lines of squares on all four sides of the board. The ratio between black and white squares remains one at all stages of this procedure and, therefore, also when considering the limit of an infinitely extended chessboard. Many alternative regularization procedures that appear equally natural yield the same result. Of course, one can regularize in a different way, by systematically including more black than white squares (or conversely) at any stage of the regularization procedure, yielding a different limiting ratio. But, as one may feel, there is still a preferred sense in which the ratio between the number of black and white squares is one and hope that the same applies to regions with different values of the parameters in the space-time framework of eternal inflation.

Unfortunately, the problem of selecting an appropriate regularization procedure is very difficult to solve for the landscape multiverse formed by island universes as produced by eternal inflation. In that setting, the different regularization procedures are called "cosmic measures," and there are many serious contenders on the table for being the right measure to be used in the landscape multiverse. The suggested ones include the following:

One of the most straightforward choices is the so-called *proper time measure*. This regularization procedure is defined by starting with an arbitrary finite space-like surface that defines an initial proper time $\tau = 0$ and then considering the space-time volume covered by time-like geodesics that emanate orthogonally into the future direction from that initial surface. Ultimately, one is interested in the limit $\tau \mapsto \infty$, where the weighting factors of regions with different parameters will be independent of the chosen initial surface.

Unfortunately, this regularization procedure yields disastrous results when applied in the context of eternal inflation because it produces probabilities that strongly disfavor finding oneself to be an observer at such a late stage of cosmic history – 13.7 billion years after the Big Bang – as we apparently are. The reason for this rather bizarre problem is that, at any given proper time τ, the vast majority of observers having been created find themselves in exceedingly young universes.

The proper time measure thus leads to a probability distribution in which older universes are exponentially suppressed in proportion to their age. As argued in [Linde et al., 1995; Guth, 2007; Bousso et al., 2008], by the standards of this measure, most observers will be quantum fluctuations that spontaneously arise shortly after the Big Bang of their respective universes. This even includes observers with brains and nervous systems exactly as our own, whose psychological states are plausibly qualitatively identical to ours. However, being the result of quantum fluctuations, the brain states of these observers do not reflect any actual histories, so the apparent memories of these observers are deceptive and tell them about a past that never took place.

A similar problem has long been known to result from a famous cosmological scenario originally suggested by Ludwig Boltzmann [1895], who credits his former assistant Schütz with the idea underlying the scenario. According to that scenario, the vast majority of space-time is in a high-entropy thermal equilibrium state. Observers only exist inside localized pockets with thermal fluctuations in which the requirements for their existence are met. As first pointed out by Eddington [1931] and further discussed in [Barrow and Tipler, 1986; Albrecht and Sorbo, 2004; Carroll, 2020], it follows from this proposal that the vast majority of observers with some given subjective state of experience are the result of entropy fluctuations as they are minimally required for the formation of such observers. Notably, the past of such observers is not lower entropy than their present, which means that their apparent records and memories are "fake" and do not reflect any actual history. Honoring Boltzmann, such pathological observers are affectionately called *Boltzmann brains*, and their cousins who dominate the cosmos created by eternal inflation according to the proper time measure have been dubbed *Boltzmann babies* [Bousso et al., 2008].

The problem with theories that entail a sizeable proportion of Boltzmann brains among observers is *not* that such theories predict us to be Boltzmann brains, whereas we are not. Rather, as pointed out by Carroll [2020], the problem is that we cannot even seriously consider the possibility that we might be Boltzmann brains (or Boltzmann babies), or at least not act in any way that takes this possibility seriously. Doing so would lead to severe cognitive dissonance in that it would require us, at each step, to radically question the reliability of all our memories and records. This would also undermine our trust in the data that motivate us to take our leading candidate cosmological theories seriously in the first place. Assigning a non-negliglible probability to being a Boltzmann brain (or baby) is thus not coherently possible. If, indeed, Boltzmann babies dominate any observer reference class formed by using the proper time measure, that measure is not a serious candidate regularization procedure for the cosmos generated by eternal inflation.

The Boltzmann baby problem is avoided by other measures that are similar to the proper time measure in spirit such as the scale-factor measure and the light cone time measure. The scale-factor measure, whose properties are thoroughly discussed by Bousso et al. [2009], is used by De Simone et al. [2008] in one of the most-discussed attempts to account for the value of the cosmological constant using the landscape multiverse. The light cone measure is investigated in detail in [Bousso, 2009].

Yet another proposal is the pocket-based measure [Garriga et al., 2006], which is – implicitly or explicitly – used in most of the earlier attempts to account for the value of the cosmological constant. However, as shown by Page [2008], it also leads to a Boltzmann brain (though not a Boltzmann *baby*) problem because it seems to entail a dominance of observers who result from thermal fluctuations in the later stages of cosmic evolution. It is unclear, however, how problematic this consequence really is because it rests somewhat delicately on the interpretation of quantum theory. Notably, that Boltzmann brain problem may be avoidable on the Everett [Boddy et al., 2017] and pilot wave [Goldstein et al., 2017] interpretations.

Finally, the causal diamond measure [Bousso, 2006] has attracted a lot of attention. It considers the intersection of the forward light cone of some point lying early on a time-like geodesic and the backward light cone lying late on the same geodesic. This measure in combination with the observer proxy *entropy production* is used in [Bousso et al., 2007] in one of the main attempts to account for the measured value of the cosmological constant.

8.3.3 Can We Determine a Uniquely Correct Measure?

Seeing this plethora of options in the choice of cosmic measure, we may ask whether that choice actually matters for which observations are predicted from the theory. The unambiguous answer to this question is "yes." For example, as recently demonstrated using state-of-the-art numerical methods [Barnes et al., 2018], leaving all other parameters fixed, the predicted value of the cosmological constant depends for its order of magnitude on the chosen cosmic measure.

When assessing different candidate measures, cosmologists consider which ones have least conceptual problems and can be defined most independently of physical assumptions, e.g., concerning the existence of "baryons" or "photons." Furthermore, they assess the candidate measures according to their perceived elegance and simplicity, and, perhaps most crucially, according to whether they lead to predictions for measurable parameters that agree with the observed values of those parameters. Effectively, cosmologists thereby treat cosmic measures analogously to scientific theories by assessing them in the light of standard theoretical virtues such as consistency, elegance, and simplicity, but also empirical adequacy. For example, Bousso et al. [2008, abstract] argue that a measure should be assessed "[l]ike any other theory" – namely, according to whether it is "simple, general, well-defined, and consistent with observation." And, indeed, according to Tegmark, the analogy between physical theories and cosmological measures goes very far: just as there supposedly is some correct physical theory that, figuratively speaking, "nature subscribes to," there is, according to him, "some correct measure that nature subscribes to" [Tegmark, 2005, p. 2].

Tegmark expresses confidence that a preferred measure can be identified in analogy to how the Liouville measure in phase space was identified as the right measure in classical statistical mechanics and the Born measure in Hilbert space quantum mechanics:

[A] measure problem (how to compute probabilities) plagued both statistical mechanics and quantum physics early on, so there is real hope that inflation too can overcome its birth pains and become a testable theory whose probability predictions are unique. [Tegmark, 2005, p. 13]

But how far the analogies between classical statistical mechanics and quantum mechanics, on the one hand, and the multiverse of eternal inflation, on the other hand, really go and to what degree Tegmark's comment is historically accurate is not clear. The closest analogy might be the one between eternal inflation and Everettian quantum theory because in both cases the objects that probabilistic weights are assigned to by the measure are indeed real. But, as will be discussed in Section 10.1, it is highly doubtful whether the Born measure can really be regarded as specifying an agent's rational self-locating credences in the Everettian quantum context. And this despite the fact that the Born measure is systematically preferred in an important sense, namely, by being dynamically preserved under unitary time evolution.

Against Tegmark, Arntzenius and Dorr [2017, section 20.7] argue that the idea of an objectively correct measure for eternal inflation – the one that, as Tegmark puts it, "nature subscribes to" – is highly problematic. To see why, it helps to realize that its proponents face the following dilemma: either facts about the correct measure supervene on other physical facts – e.g., facts about what the correct physical theory is, facts about the overall structure of space-time, and facts about the distribution of matter and energy – or they do not supervene on such facts.

If facts about which measure is correct do *not* supervene on such other facts, the correct measure is a separate matter, which means that any specific complete cosmic history – complete in terms of overall matter and energy distribution – does not uniquely determine the correct cosmic measure. At least in principle, it must be compatible with at least two, possibly more, cosmic measures. But which of those measures is the one that nature subscribes to would then be unknowable to observers because observations inside the cosmos could not possibly distinguish between them. Why would we stipulate facts about measures if those facts are, in principle, epistemically inaccessible? Postulating a correct cosmic measure that does not supervene on cosmic history does not seem to be a coherent option.

If, on the other hand, the correct measure *does* supervene on the other physical facts, what we are actually trying to determine when competing different measures against each other are structural aspects of those other, measure-determining facts. Empirically establishing the correct measure would, in this case, amount to determining how the measure-determining physical facts lie – e.g., what the overall space-time structure or matter-energy distribution of the cosmos might be like. I am not aware of any serious attempts to establish some specific cosmic measure as systematically privileged in the context of eternal inflation in the sense in which the Born measure is privileged in Everettian quantum theory.

It should also be noted that even if some specific measure were established as physically privileged in the context of eternal inflation, this would not by it-self show that this measure should guide our assignment or probabilities. The example of the Born measure in Everettian quantum theory shows this. As I argue in Section 10.1, the fact that it is physically special does not suffice to show that the Born rule, which is based on it, delivers rational self-locating credences in an Everettian world.

These difficulties may make the idea mentioned before very tempting to simply treat the "correct" measure in analogy to the correct cosmological theory as something to be "tested" by comparing the predictions based on them with respect to empirical adequacy, as if they were physical theories in their own right. This approach is immediately suspicious, however, because if one uses empirical adequacy as a criterion of choice between measures, one dramatically reduces its impact in the choice between theories. In what follows, I argue that cosmic measure and observer proxy, when adjusted in the light of predictions that are not confirmed, function as *researcher degrees of belief* in a highly problematic way that makes any predictions derived from the multiverse theories by means of their help completely unreliable.

8.4 Researcher Degrees of Freedom

As we have seen, predictions derived from specific multiverse theories depend strongly on the chosen observer proxy and the chosen cosmic measure. Accordingly, whether certain observations will be regarded as confirmatory of some multiverse theory or as in conflict with it will, in general, also depend heavily on the chosen observer proxy and measure. But this creates a difficulty for multiverse theory testing that parallels a problem that seems to have significantly contributed to the so-called *replicability crisis* in the social sciences. As attempts to ascertain the robustness of various published outcomes of a large number of studies in the social sciences have shown, many of these outcomes failed to replicate when the original studies were performed a second time. This finding also sheds doubt on the claimed results of many other studies in the social sciences for which hitherto no replication attempts have been made.

The causes of the replicability crisis in the social sciences are numerous and complex, but it is widely believed that one phenomenon in particular has contributed significantly to it – namely, scientists' (often unintended) exploitation of so-called *researcher degrees of freedom*. This concept, which will also help us understand why exactly predictions derived from multiverse theories using observer proxies and cosmic measures are notoriously unreliable, is elucidated in the following passage of an influential paper by social scientists Simmons et al. [2011]:

The culprit is a construct we refer to as researcher degrees of freedom. In the course of collecting and analyzing data, researchers have many decisions to make: Should more data be collected? Should some observations be excluded? Which conditions should be combined and which ones compared?

Which control variables should be considered? Should specific measures be combined or transformed or both? It is rare, and sometimes impractical, for researchers to make all these decisions beforehand. Rather, it is common (and accepted practice) for researchers to explore various analytic alternatives, to search for a combination that yields "statistical significance," and to then report only what "worked." [Simmons et al., 2011, p. 1359]

A positive result counts as "statistically significant" if, assuming that it does *not* reflect a genuine feature of the population studied, the probability of obtaining it as a result of mere chance in the sampling process is low. In principle, the requirement that results be statistically significant is intended to make the process of scientific hypothesis testing robust and reliable. But, as outlined in the passage just quoted, actual practices of data collection and analysis in the social sciences make it easy for researchers who – intentionally or unintentionally – *want* to see some suggested hypothesis come out confirmed to end up with statistically significant results in favor of that hypothesis, whether or not it is actually true.

Whether they consciously intend to or not, researchers can increase their chances of obtaining significant results if they exploit various researcher degrees of freedom that they have at numerous points in the procedures of data collection and data analysis. According to Simmons et al., the most important researcher degrees of freedom include amount of data collected, inclusion or dismissal of specific data points (e.g., "outliers"), and combination of the data points obtained in specific subgroups, and there are likely many more. Typically, for at least some ways of exploiting the researcher degrees of freedom, the hypothesis favored by the researchers will come out as confirmed at a certain level of statistical significance.

Ideally, researchers will pay attention to the pitfalls of possible researcher degrees of freedom and take active steps to avoid them. If they do not, the predictable result is a proliferation of statistically significant results that do not replicate when subjected to renewed empirical scrutiny – exactly what has been found in the past few years in the social sciences.

Returning to cosmology, since our empirical access is confined to our own universe, there can be no repeated universe sampling, unlike in the social sciences, even though we can, of course, remeasure the values of parameters that we have already measured. As a consequence, if our methods of extracting empirical predictions from multiverse theories are unreliable, we cannot hope that a replicability crisis will emerge and make us aware of their unreliability. To prevent getting stuck in permanent error, we must scrupulously identify any aspects of those methods that make them susceptible to failure.

And, indeed, on reflection, it seems hard to avoid the impression that, in attempts to make multiverse theories testable along the lines described earlier, observer proxies and cosmic measures effectively function as researcher degrees of freedom in the sense of Simmons et al.: the choices of both are not rigorously constrained by widely accepted criteria that have a credible claim to objectivity, yet both have an extremely large influence on the outcomes of the predictions made.

Moreover, researchers openly admit that they make those choices with an eye on deriving predictions that are confirmed by observations. Notably, the idea of *testing* cosmic measures, as explicitly advocated by Tegmark [2005], Linde [2007], and Bousso et al. [2008], amounts to openly using the measure as a researcher degree of freedom in that one freely adjusts the measure used in order to derive predictions that reproduce the observations made as accurately as possible. (Smeenk [2014] makes essentially the same point without using the notion of researcher degrees of freedom.)

To highlight that these are not merely abstract, theoretical worries, it is useful to note that derivations like the one by Bousso et al. [2007] of which value of the cosmological constant we should empirically expect from the landscape multiverse scenario contain little, if any, appeal to specific features of string theory, even though, technically, string theory is, of course, a central ingredient of the landscape multiverse scenario. Only rather generic assumptions about the abundance of vacua with values of the cosmological constant not many orders of magnitude different from zero are used.

Indeed, the main use of "predictions" of the value of the cosmological constant in a multiverse setting seems to have been to discriminate between distinct measures. There are no concrete prospects for shifting the focus from measure competition to theory competition.

Moreover, even if the problem of measure choice is set aside, unless careful attempts are made to connect the choice of observer proxy to the observer reference class mandated by the assumed background knowledge D_0 in the light of the BIC, the observer proxy can be expected to function as a researcher degree of freedom in the same way. Thus, unless advocates of the landscape multiverse find a way to keep *both* those researcher degrees of freedom firmly in check so that their choice cannot possibly be tailored to deliver specific outcomes, empirical predictions derived from the landscape multiverse cannot be trusted.

8.5 Why Testing Multiverse Theories Is So Damn Difficult

There may well be multiverse theories that are not plagued by the infinities that appear in eternal inflation and haunt the landscape multiverse. Such multiverse theories would not give rise to any version of the cosmological measure problem. But even for those theories, attempts to extract concrete empirical predictions from them using self-locating belief theory might still be plagued by researcher degrees of freedom such as the observer proxy. Those would continue to make it extremely difficult to derive reliable predictions from such theories.

In fact, one should expect further problematic researcher degrees of freedom to arise beyond those already discussed.

A candidate example is the decision to include – or not include – the measured value of a certain parameter λ' in one's background knowledge D_0 when assessing the likelihood of finding $[\lambda]$, given the theory T at issue. Alternatively, one can treat the value of that λ' as a quantity to be predicted as well by that multiverse theory; i.e., one could consider the likelihood of finding specific combinations of values of λ and λ', given somewhat less rich background knowledge D_0', which leaves open the value of λ'.

For instance, as shown by Tegmark and Rees [1998], treating the relative amplitude Q of primordial inhomogeneities as variable rather than as fixed at its measured value, as was done by Martel et al. [1998], has a significant effect on the typical value of the cosmological constant Λ to be found by observers. This is not per se problematic, for if one includes the measured value of Q in the background knowledge D_0, this will have ramifications for the reference class of observers that one should use according to the BIC and for the prior $P(T|D_0)$ that one assigns to the multiverse theory T at issue: switching from background knowledge that includes the value of Q to background knowledge that does not include it requires an adjustment of both the observer reference class used and the prior assigned to T.

In principle, such a reference class switch is perfectly possible and entirely unproblematic. If the reference class is always adjusted appropriately as the considered background knowledge is changed, the rational posterior probability assigned to any candidate theory T will not depend on it.

For example, if D_0 is some background knowledge that does not take into account the known value of Q and D_0' is some background knowledge that does take into account that value, then $P(T|Q, D_0)$ and $P(T|D_0')$ should be the same. The impact of obtaining information about the value of Q would have been taken into account properly by conditioning with respect to it. But it is doubtful whether shifts in background knowledge are always properly tracked *in practice*. Usually, no detailed account of what is contained in the background knowledge is provided. As a consequence, the choice of which parameters one treats as fixed and which as variable plays the role of yet another researcher degree of freedom.

There are probably even more choices that effectively function as researcher degrees of freedom in multiverse theory testing – for example, the choice of whether and, if so, how to take into account cosmic inhomogeneities and anisotropies. That those have a potentially big impact on predictions of the value of the cosmological constant has been demonstrated by Phillips and Albrecht [2011]. Whether and, if so, how inhomogeneities and anisotropies are modeled can thus be expected to play the role of a further important researcher degree of freeom.

There is no reason to believe that the list of easily – and unintentionally – exploitable researcher degrees of freedom is thereby complete. Given the human susceptibility to confirmation bias, those will all inevitably impair the credibility of claimed empirical predictions from multiverse theories.

8.6 A Way Out à la "Open Science"?

The diagnosis of the problems that arise in attempts to make multiverse theories testable as having to do with "researcher degrees of freedom" is taken over from the analysis of the replicability crisis of the social sciences. Since that analysis was proposed, a significant number of steps and initiatives have been proposed to combat those replicability problems.

Can we perhaps transfer those steps and initiatives to the context of multiverse cosmology; i.e., can the problem of researcher degrees of freedom in testing multiverse theories perhaps be addressed in a similar way to how many concerned social scientists now try to solve the replicability crisis in their fields? Key suggested strategies, which are currently tried, to improve the robustness of findings in the social sciences and thereby clarify which claimed discoveries are reliable and which not include the following:

- Preregistration of studies: In this new publication format, scientists lay out their research idea together with a plan of data collection and analysis prior to actually carrying out the research. Journals provisionally commit to publication of the article on the research, irrespective of whether the outcomes that are obtained end up confirming or disconfirming the initial hypothesis.

This approach is intended to reduce incentives for – intentionally or unintentionally – producing results that support the research hypothesis. It is hoped to thereby minimize the detrimental effect of confirmation bias.

- Increased publication of null results and replications, including failed ones, in prestigious venues: This measure is also intended to decrease the pressure on scientists to produce results that confirm specific theoretical ideas to which they have committed themselves.

- Redefinition of statistical significance: Some researchers suggest that a result should only be acknowledged as "statistically significant" if deviations from what one would expect based on the "null hypothesis" (absence of the suggested effect) are more pronounced than usually required until now. Specifically, they propose that the conventional statistical significance "level" should be reduced from 5% to the more ambitious 1%. This is supposed to decrease the danger of publication of "false positives" (which, in reality, seems larger than the danger of publishing "false negatives"). Recently, some researchers [Amrhein et al., 2019], supported by more than 800 signatories, have even called for an end of the central role that the concept of statistical significance conventionally has in contemporary quantitative science.

- Improve the understanding of statistics: The very meanings of key statistical concepts are often unclear to scientists. Indeed, the proper understanding of conventional (frequentist) statistics is quite intricate and in many respects not at all straightforward. For instance, ascriptions of probability to scientific hypotheses, as they are routinely made in this book, are illegitimate in its contexts.

Failure to be completely aware of the meaning of key statistical concepts can make scientists unaware of situations in which standard techniques of data collection and analysis can give rise to fallacies. Better educating scientists about the meanings of statistical concepts is therefore one promising step toward improving the robustness and reliability of the presented findings. Concretely, making scientists aware of the concept of researcher degrees of freedom and how these arise is part of this education.

These strategies can, to some degree, be transferred, at least in spirit, to attempts of extracting concrete predictions from multiverse theories in cosmology. Notably, as this book tries to do, one can make cosmologists aware of the problem of researcher degrees of freedom and how these may arise in their work.

An analog of a more ambitious definition of statistical significance that applies to multiverse cosmology could be imposing the requirement that derivations of quantities to be expected in multiverse theories should be measure independent, observer proxy independent, etc. It seems doubtful whether this goal can realistically be achieved for interesting and nontrivial derivations. Rigorously implementing this requirement might make the derivation of empirical consequences from specific multiverse theories downright impossible.

Implementing an analog of preregistration in multiverse theory testing might also be difficult. Typically, cosmologists do not derive any predictions from multiverse theories for quantities to be measured or data yet to be taken. Rather, they try to reproduce the already-known values of quantities such as the cosmological constant. This makes it hard to imagine how one could implement anything resembling pregistration in multiverse cosmology.

I conclude that the prospects for curing the predictivity crisis in multiverse cosmology – that is, the prospects for extracting definite empirical predictions (or postdictions) from multiverse theories – are far worse than the prospects for overcoming the replicability crisis in the social sciences. The chances of ever making reliable empirical predictions based on concrete multiverse theories using the anthropic approach outlined in the previous chapter and the first section of this chapter are bad.

In the chapters of Part I of this book, we have encountered good reasons to take the possibility that we might live in a multiverse seriously, at least in principle, and perhaps even consider it a live possibility. Against that background, the moral of this chapter is particularly disappointing. While we should not discard multiverse theories as unworthy of our attention, obtaining compelling empirical evidence for any concrete multiverse theory – even if that theory happens to be true! – seems, for the foreseeable future, out of reach.

9

Puzzles of Self-Locating Belief

As we have seen in the last chapter, the idea of testing multiverse theories by using self-locating indifference runs into formidable practical difficulties because of various researcher degrees of freedom that researchers may – often unintentionally – exploit. But self-locating indifference may even run into far more basic *systematic* difficulties in that it leads to unacceptable verdicts on various intriguing *puzzles of self-locating belief*. Famous such puzzles include the *Sleeping Beauty* problem, the *Doomsday Argument*, and the *Presumptuous Philosopher* problem.

The present chapter investigates those problems and the difficulties that they potentially pose for self-locating indifference. The verdict will be that the puzzles of self-locating belief can be resolved or, to the degree that they remain puzzling, that they do not indicate any problems for self-locating indifference specifically. Thus, puzzles of self-locating belief do not pose any *additional* challenges to the avenue toward testing multiverse theories discussed in the previous two chapters beyond the challenges identified there. They are fascinating in themselves, however, and are therefore worth a chapter in their own right, even if they do not ultimately pose an additional fundamental challenge to multiverse theory testing.

9.1 Troubles for Self-Locating Indifference?

9.1.1 The Doomsday Argument

A puzzle of self-locating belief where self-locating indifference may seem to lead to counterintuitive conclusions is the notorious *Doomsday Argument*. In its simplest version, the Doomsday Argument involves two competing hypotheses H_1 and H_2 that differ on the total number of humans ever to exist – namely, either N_1 and N_2. (In this simple version of the Doomsday Argument, it is assumed that all other hypotheses are excluded beforehand.) Using numbers borrowed from Bostrom [2001], either $N_1 = 200$ billion or $N_2 = 200$ trillion humans are going to have lived. The hypothesis H_1 is a "doom soon" hypothesis because it entails that the number of humans yet to be born is rather small.

Let us assume that our empirical evidence suggests an optimistic assessment: it is much more probable that there will be 200 trillion people rather than merely 200 billion, so we say, for example, $P(H_1) = 0.05$ and $P(H_2) = 0.95$. Next, let us assume that you learn that you are the $n = 60$ billionth human to exist and that there are no other observers beside humans that we need to include in our reference class. This can be interpreted as a (temporally) self-locating insight, and it seems natural to use self-locating indifference with respect to the reference class of humans ever to be born to study its evidential impact. If we do so, it leads to the ratio of posterior probabilities $P^+(H_1)/P^+(H_2)$:

$$\begin{aligned} \frac{P^+(H_1)}{P^+(H_2)} &= \frac{P(H_1|n)}{P(H_2|n)} \\ &= \frac{P(n|H_1)P(H_1)}{P(n|H_2)P(H_2)} \\ &= \frac{N_2}{N_1} \cdot \frac{P(H_1)}{P(H_2)}. \\ &= 1000 \cdot \frac{0.05}{0.95} \\ &\approx 50, \end{aligned} \tag{9.1}$$

where self-locating indifference (the self-sampling assumption)

$$P(n|H_i) = \frac{1}{N_i}, \tag{9.2}$$

for $i = 1, 2$ was used in the transition from the second to the third line of Eq. (9.1), together with the fact that $n < N_1$ and $n < N_2$.

The startling conclusion of this reasoning is that you should expect H_1 ("doom soon") to be true even if the input probability $P(H_1)$ has been substantially lower than the input probability $Pr(H_2)$. This seems counterintuitive: learning where in human history one was born should not, intuitively, evidentially support the theory that the end of humanity is near. Self-locating indifference is invoked in a crucial step in the Doomsday Argument, so – given this argument's implausible conclusion – it seems natural to regard it as showing self-locating indifference to be problematic.

9.1.2 The Appearance of Anomalous Causal Powers

Before discussing whether this is an attractive diagnosis, let me mention an even weirder aspect of reasoning of the type used in the Doomsday Argument: as

Bostrom points out, such reasoning seems to sometimes recommend acting as if one had what he calls "anomalous causal powers" [Bostrom, 2001, p. 368]. He proposes some scenarios where this happens, which are incidentally not among the most-discussed problems of self-locating belief in the literature: the three *Adam and Eve experiments* and the *UN++-Gedanken experiment*. The story that Bostrom calls *Lazy Adam* is particularly suitable to highlight what he has in mind:

> Assume ... that Adam and Eve were once the only people and that they know for certain that if they have a child they will be driven out of Eden and will have billions of descendants. ... [T]hey have a foolproof way of generating a child, perhaps using advanced *in vitro* fertilization. Adam is tired of getting up every morning to go hunting. Together with Eve, he devises the following scheme: *They form the firm intention that unless a wounded deer limps by their cave, they will have a child.* Adam can then put his feet up and rationally expect with near certainty that a wounded deer – an easy target for his spear – will soon stroll by. [Bostrom, 2001, p. 367]

Adam's and Eve's reasoning seems bizarre, but, just as for the Doomsday Argument, it is surprisingly difficult to determine what, if anything, is wrong with it. We can formalize it as follows: let H_1 be the hypothesis that some wounded deer will turn up, which means that Adam and Eve will refrain from having children and, accordingly, will remain the only humans ever to exist. So, the total number of observers ever to exist according to H_1 is $N_1 = 2$. Next, let H_2 be the alternative hypothesis that no wounded deer will turn up such that Adam and Eve, following their firm intention, will have children so that, at the end of the world many thousands, millions, or billions of years later, a large number – say, $N_2 = 10^9$ – of observers will have existed.

Now let us allow Adam to use his knowledge that he is the first observer ever to have lived ($n = 1$), which results in

$$\frac{P^+(H_1)}{P^+(H_2)} = \frac{P(H_1|n=1)}{P(H_2|n=1)}$$
$$= \frac{P(n=1|H_1)P(H_1)}{P(n=1|H_2)P(H_2)}$$
$$= \frac{N_2}{N_1} \cdot \frac{P(H_1)}{P(H_2)}, \qquad (9.3)$$

where the first line uses Bayes's theorem and the second uses Eq. (9.2) together with the fact that $n = 1$ is compatible with both H_1 and H_2. Assuming Bayesian conditioning and that Adam has a very small prior credence that a wounded deer will turn up – say, $P(H_1) = 10^{-7}$ (which means $P(H_2) = 1 - 10^{-7}$) – we obtain for his rational *posteriors*

$$\frac{P(H_1|n=1)}{P(H_2|n=1)} = \frac{10^9}{2} \cdot \frac{10^{-7}}{1-10^{-7}} \approx 50. \qquad (9.4)$$

So, as Bostrom suggests, if he uses his knowledge that he is the first human ever to have lived in the way just outlined, Adam will be confident that a wounded deer will walk by. The conclusion persists qualitatively even if a prior much smaller than $P(H_1) = 10^{-7}$ is used. Even if Adam is then no longer confident that a wounded deer will turn up, he will still seem overly optimistic given how farfetched the possibility really is.

As Bostrom notes, this verdict on Adam's rational posterior credences is highly counterintuitive:

We ... have ... the appearance of psychokinesis. If the example works, which it does if we assume SSA [i.e., Eq. (9.2)], it almost seems as if Adam is causing a wounded deer to walk by. For how else could one explain the coincidence? Adam knows that he can repeat the procedure morning after morning and that he should expect a deer to appear each time. Some mornings he may not form the relevant intention and on those mornings no deer turns up. It seems too good to be mere chance; Adam is tempted to think he has magical powers. [Bostrom, 2001, p. 367]

Beside *Lazy Adam*, Bostrom discusses two other Adam and Eve experiments, *Serpent's Advice* and *Eve's Card Trick*, which involve apparent anomalous causal powers in similar ways: in Serpent's Advice, these powers have the flavor of "anomalous precognition" [Bostrom, 2001, p. 367], and in Eve's Card Trick, they appear as apparent anomalous "backward causation" [Bostrom, 2001, p. 368]. As this latter example shows, the apparent anomalous causal powers need not be forward directed in time.

According to Bostrom, strange though Adam's credences seem, they look less unacceptable if we realize that an important bit of evidence that *we* have – namely, that there will be many other observers besides Adam and Eve, including us – is simply unavailable to Adam. Bostrom nevertheless concedes that the recommendation that Adam should really have credences that conform to Eq. (9.4) is "deeply counterintuitive" [Bostrom, 2002, p. 157]. But he also points out that accepting them does not mean to ascribe *real* anomalous powers to Adam and Eve even if we endorse their counterintuitive reasoning as rational after all: "There is ... no reason to ascribe anomalous causal powers to Adam. Eve and Adam would rationally think otherwise, but they would simply be mistaken" [Bostrom, 2001, p. 373].

Before discussing possible ways to avoid Eve and Adam's counterintuitive reasoning, it is useful to point out – as Bostrom may not have realized – that one can also set up the Doomsday Argument in such a way that it involves apparent anomalous causal powers. To do so, assume that whether H_1 or H_2 turns out to hold in the Doomsday Argument depends on the success of a group of terrorists, who are trying to construct a pernicious machine that, if completed, would put an immediate end to humanity (and, so, make H_1 true). Fortunately, constructing this machine is difficult, and the objective chance of the terrorists to succeed is a meager

$P(H_1) = 0.05$. (We have some experience with the construction of machines that are of the same type but less pernicious, which allows us to assign this probability). If they succeed, $N_1 = 200$ billion humans will have lived; if not, $N_2 = 200$ trillion. Based on the information that you are the 60 billionth human being to be born, and using the same reasoning as in Eq. (9.1), you should have credence $P(H_1|n) \approx 0.98$ that the terrorists will succeed in their work.

Earlier work on the Doomsday Argument (e.g., by Leslie [1989], who endorses it) highlights that, if the argument is valid, we should take any factors that may cause humanity to go extinct much more seriously than we would otherwise do. Looking at the argument through the lens of apparent anomalous causal powers, this warning translates into the recommendation that we should treat people trying to bring about the end of humanity – e.g., the terrorists – as if having *enhanced* causal powers and people trying to preserve humanity as if having *reduced* causal powers.

To illustrate how odd this conclusion is, consider some nerdy enthusiast of the type of machine that the terrorists try to construct. According to the Doomsday Argument, if he cares more about contributing to the successful construction of such a machine than about humanity's future, joining the terrorists is an excellent strategy for him to achieve his aims – even if he could collaborate with more-skilled collaborators when constructing the machine for neutral, perhaps even humanity-preserving, purposes. This recommendation seems particularly difficult to accept.

9.2 Reacting to the Puzzles

9.2.1 A Panorama of Responses to the Doomsday Argument

The Doomsday Argument has many critics. There are others besides Leslie who, partly with reservations, defend it and endorse its conclusion – notably Pisaturo [2009], Lewis [2010], and Bradley [2012], who argue (along different lines) that the conclusion is only apparently as implausible as it initially seems and only when viewed through the lens of the distorting and misleading characterization of H_1 as "doom soon." As these authors point out, when one learns that one's birthrank is so low as to be compatible with the truth of H_1, this actually *raises* the expected number of future humans because it entails that *if* H_2 is true, that number will be large. Thus, according to the Doomsday Argument, while learning one's (early) birthrank supports the hypothesis according to which humanity's existence is comparatively short, it simultaneously increases the expected future duration of humanity's future.

But the appearance of (apparent) anomalous causal powers entailed by the Doomsday Argument is not made any more palatable by these considerations. It therefore seems doubtful whether we should be content with them.

As mentioned in the introduction to this chapter, the Doomsday Argument is sometimes presented as a potential problem for self-locating indifference Eq. (9.2), which is indeed centrally used in the derivation Eq. (9.1) of the Doomsday Argument. In principle, one can indeed evade the Doomsday Argument's conclusion by giving up self-locating indifference. But unless one chooses a conditional prior $P(n|H_2)$, which – when considered as a function of n – is highly peaked around $n = 1, 2$ (which is necessary to have $P(n = 1|H_1) \approx P(n = 1|H_2)$), the effect that the ratio of the posteriors $P(H_1|n)/P(H_2|n)$ differs strongly from the ratio of the priors $P(H_1)/P(H_2)$ will persist, and this will suffice to reproduce Adam's conclusion and the Doomsday Argument in their qualitative features.

Moreover, in a hypothetical situation where one knows H_2 to be true – i.e., that there are in total $N_2 = 10^9$ observers, but where one has not the faintest idea who among them one is – there is just no reason to assume with near certainty that one will be among the very first two ever to exist (as $P(n = 1|H_1) \approx P(n = 1|H_2)$ would require). To conclude, it is difficult to see how one might justify evaluating $P(n|H_2)$ in a manner sufficiently different from Eq. (9.2) to avoid the conclusion reached by Adam in its qualitative features.

Another strategy for avoiding the Doomsday Argument is to regard its conclusion as the artifact of an arbitrary choice of observer reference class that includes only humans. The discussions by Eckhardt [1993] and Neal [2006] can be read as versions of this idea. As Neal puts it:

I take [the Doomsday Argument] to be absurd, primarily because the answer it produces depends arbitrarily on the choice of reference class. Bostrom (2002) argues that this choice is analogous to a choice of prior in Bayesian inference, which many are untroubled by. However, a Bayesian prior reflects beliefs about the world. A choice of reference class has no connection to factual beliefs, but instead reflects an ethical judgement, if it reflects anything. It is thus unreasonable for such a choice to influence our beliefs about facts of the world. [Neal, 2006, p. 3]

Even if one does not concur with Neal that the choice of reference class is "ethical," it is certainly true that choosing a reference class of "humans" is, in many respects, arbitrary. Neither to the past nor perhaps to the future does this reference class have an obvious sharp boundary, and even inasmuch as it has a sharp boundary the focus on humans rather than, say, primates or more broadly defined intelligent agents seems questionable to begin with.

As promising as this attack on the Doomsday Argument may initially seem, it is arguably not very attractive. Adam-style reasoning and Doomsday-style reasoning seem questionable *per se*, not merely for the particular reference classes used. As soon as one chooses a fixed reference class and does not change it when updating with respect to the self-locating information at issue, a Doomsday-style conclusion can be derived. There is every reason to expect that we would judge the resulting doomsday shift every bit as implausible as we do in the original Doomsday Argument.

Another solution that focuses on the observer reference class is offered by Bostrom [2002]. There, Bostrom develops an account of self-locating belief according to which the observer reference class *is* changed when self-locating information comes in and where this is done in such a way that the credences with respect to the two hypotheses H_1 and H_2 remain constant: $P^+(H_1) = P(H_1)$ and $P^+(H_2) = P(H_2)$. This suggestion has counterintuitive consequences, however, which are spelled out in Section 9.3.1, when addressing the so-called *double-halfer* solution to the Sleeping Beauty problem.

Even more radically, Norton [2010] diagnoses the Doomsday Argument as exposing a shortcoming of the probabilistic Bayesian methodology used to derive it and proposes to abandon it altogether. According to the Bayesian analysis, having birthrank $n < N$ supports H_1 to the degree that it disfavors H_1's negation H_2. It does the latter because if H_2 had been true, having a higher birthrank than N would have been likely. But according to Norton, learning of one's low birthrank $n < N$ is evidentially neutral between the two hypotheses and, therefore, should not have any such effect. He proposes an alternative formalism of "completely neutral support" (CNS) instead, in which the derivation of the Doomsday Argument is blocked.

Norton's criticism of Bayesianism applies to the "self-locating" probabilities assigned to self-locating indices such as birthrank, conditional on hypotheses. According to him, one should simply not reason in terms of probabilities such as $P(n|H_1)$ and $P(n|H_2)$, let alone base any substantive conclusions on such reasoning. This criticism translates into a general attack on the use of something like the "xerographic distributions" introduced in Chapter 7. It would also undermine the new fine-tuning argument for the multiverse presented in Chapter 6.

The price to pay for accepting Norton's response to the Doomsday Argument would thus include rejecting the framework for testing multiverse theories outlined in the previous section and the new fine-tuning argument. Norton himself is highly critical of multiverse theories and pessimistic about their testability, so this does not trouble him. But for adherents to the strategies to test multiverse theories discussed in the previous two chapters, this route is clearly not an option.

In response to Norton's general worries concerning how Bayesianism handles evidence that does not allow one to determinately discern competing hypotheses, Bénétreau-Dupin [2015b] suggests a framework of imprecise probabilities instead of Norton's inductive logic framework.

In Bénétreau-Dupin [2015a], he applies this framework to refute the Doomsday Argument by focusing on a perhaps more realistic version, where one considers a wider range of theories concerning the total number of humans ever to have lived. In the absence of strong reasons to assign specific prior probabilities to these theories, we should, according to him, use imprecise probabilities to express our initial uncertainty concerning which of them is correct. As he shows, this move prevents the conclusion of the Doomsday Argument from being derivable.

This achievement, however, comes again at the cost of dispensing altogether with the standard formalism of Bayesian inference, which involves sharp probabilities. As Bénétreau-Dupin highlights, his formalism also invalidates anthropic reasoning involved in multiverse cosmology as outlined in Chapter 7. Thus, for those who intend to use the Bayesian formalism as applied to problems of self-locating belief in the context of testing multiverse theories, Bénétreau-Dupin's response to Norton's challenge is therefore not an option.

9.2.2 A Simple Response in Line with the BIC

Interestingly, a much less radical option is available for rejecting Adam's reasoning and the Doomsday Argument, championed by Dieks [2007]. Dieks's simple idea is to arrive at a different numerical evaluation of the expressions used there by identifying the *input probabilities* ($Pr(H_1) = 0.05$ and $Pr(H_2) = 0.95$ in our example) with the *posteriors* $P(H_1|n)$ and $P(H_2|n)$ rather than the *priors* $P(H_1)$ and $P(H_2)$. In this solution, the posteriors favor a long future for humanity, just as they intuitively should. When applied to Lazy Adam, Dieks's reasoning yields posteriors $P(H_1|n = 1) = Pr(H_1) = 10^{-7}$ and $P(H_2|n = 1) = Pr(H_2) = 1 - 10^{-7}$, according to which – as seems plausible – Adam should not expect any wounded deer to turn up.

An interesting feature of Dieks's proposal is that, as long as one still accepts the reasoning encoded in Eq. (9.1), it leads to priors that very strongly favour the hypothesis with more observers over the one with less. This aspect is central to a first strategy of motivating Dieks's move, which derives from earlier work by him [Dieks, 1992] and is championed by Olum [2002]. This strategy invokes what Bostrom calls the *self-indication assumption* (SIA), which dictates that priors be assigned to hypotheses in proportion to the number of observers whose existence the hypotheses entail:

$$\frac{P(H_1)}{P(H_2)} = \frac{N_1}{N_2} \cdot \frac{Pr(H_1)}{Pr(H_2)}. \tag{9.5}$$

When combined with appeal to the SSA (Eq. (9.2)), this reproduces Dieks's verdict. The main problem with this strategy is that there seems to be little independent reason besides avoiding a Doomsday-style conclusion for assigning priors to hypotheses proportionally to the number of observers whose existence they entail.

Neal [2006] considers a response to the Doomsday Argument and similarly structured problems of self-locating belief that leads to recommendations in line with the SSA and the SIA combined. It involves hypothetical *companion observers* distributed uniformly across time and space. If such hypothetical companion observers found themselves coexisting with humans, without knowing whether they themselves exist before or after the potential doom event, they would take our presence as evidence in favor of a longer history, for if human history had been very short, it would have been less probable for them to be temporally located in the historical span in which humans exist.

I see three problems with this proposal: first, it is not clear why the dynamics of the rational credences of such hypothetical companion observers should be relevant to our rational credences. Second, it is unclear why those dynamics should be pictured as involving a stage in which the companion observers are disoriented about whether they exist before or after the potential doom moment. And third, the spatiotemporal distribution of the imagined companion observers enters as an additional degree of freedom in the analysis; it is not obvious why a distribution with uniform spatiotemporal density should be preferred over another one, perhaps one that more directly relates to the population density of humans in time.

Dieks himself, in his later work [Dieks, 2007], has a strategy for motivating his proposed response to the Doomsday Argument that no longer relies on the SIA but is in line with the background information constraint BIC: that the input probabilities $Pr(H_1)$ and $Pr(H_2)$ correspond to our rational credences in a situation where our background information constrains our birthrank to one that is compatible with the truth of both H_1 and H_2 – i.e., one that lies before the potential catastrophe that wipes out humanity.

In the Doomsday Argument, the situation where $Pr(H_1)$ and $Pr(H_2)$ are reasonable credences based on the available background evidence is therefore *not* one where we have completely lost our knowledge about where in the history of humanity we live and, for all that we kno, might very possibly live in a future thousands of generations later than we actually do. Consequently, it is not one where we should condition on information about our birthrank, as in Eq. (9.1), as if we had had no prior knowledge of it. Likewise, the implausible appearance of anomalous causal powers in the Doomsayer's credences can be seen as a symptom of failure to properly apply the background information constraint BIC.

9.2.3 The Presumptuous Philosopher

The main reason why Dieks's proposal is not universally accepted – which it arguably should be – is that, whether or not it is motivated based on explicit appeal to the SIA, it leads to priors $P(H_1)$ and $P(H_2)$, which systematically favor hypotheses with more observers. When understood as a general principle of rational thought, this recommendation seems unacceptable, as highlighted by Bostrom's *Presumptuous Philosopher* scenario [Bostrom, 2002, p. 124]:

It is the year 2100 and physicists have narrowed down the search for a theory of everything to only two remaining plausible candidate theories, T_1 and T_2 (using considerations from super-duper symmetry). According to T_1 the world is very, very big but finite and there are a total of a trillion trillion observers in the cosmos. According to T_2, the world is very, very, very big but finite and there are a trillion trillion trillion observers. The super-duper symmetry considerations are indifferent between these two theories. Physicists are preparing a simple experiment that will falsify one of the theories. Enter the presumptuous philosopher: "Hey guys, it is completely unnecessary for you to do the experiment, because I can already show to you that T_2 is about a trillion times more likely to be true than T_1 !" (Whereupon the presumptuous philosopher explains the Self-Indication Assumption.) [Bostrom, 2002, p. 124]

The philosopher's claim seems presumptuous, and following his recommendation appears to be irrational. Generally speaking, given "non-anthropic" priors of similar size for two hypotheses, it does not seem rational to prefer one over the other with near certainty just on grounds that it postulates vastly more observers. This, however, is what the principle SIA bluntly recommends.

In the story of the presumptuous philosophers, the background information based on which the theories T_1 and T_2 seem equally likely are only hinted at. For a solution of the problem based on the BIC, however, one would have to make that background knowledge somewhat more explicit. The solution itself may depend on the details of how this plays out in different versions of the problem.

To see this, consider first a version of the Presumptuous Philosopher scenario where T_2 differs from T_1 in that, according to T_2, the universe according to T_1 exists in a trillion causally isolated copies (or, for philosophers who believe that universes that are exact copies of each other are actually identical, approximate copies), such that it is physically impossible for the presumptuous philosopher to determine in which of the copies she is. In that case, the presumptuous philosopher's background information based on which she deems the theories T_1 and T_2 roughly equally likely $(Pr(T_1) \approx Pr(T_2) \approx 1/2)$ does not entail in which of the trillion universe copies she exists if T_2 is correct. In this case, her priors $P(T_1)$ and $P(T_2)$ correspond to the input probabilities $Pr(T_1) \approx Pr(T_2) \approx 1/2$ and, thus, do not strongly favor any of the two theories. Because the presumptuous philosopher will never be able to determine her "universe index" (the supposed analog of birthrank) and condition on it, she will never transition from those priors to extreme posteriors, which would – in this case – strongly favor the theory that entails fewer observers. Paradoxical conclusions are avoided.

Matters are different in versions of the Presumptuous Philosopher scenario where it *is* in some sense possible for the presumptuous philosopher to determine some self-locating index that plays the role of "birthrank" in the Doomsday Argument. In fact, one can set up the Presumptuous Philosopher problem as a cosmic Doomsday-type problem. In this version of the problem, the class of observers that exist according to T_1 is qualitatively identical to a subclass of the class of observers that exist according to T_2, while the vast majority of observers that exist according to T_2 (all those that are not in the subclass) are temporal and/or causal descendants of those in the subclass. This constellation is realized in a scenario where T_1 and T_2 agree on the history of the world up to some time t, such that, according to T_1, there are no observers after t, whereas, according to T_2, the vast majority of observers live after t, with their history being influenced by those who live before t.

Once the Presumptuous Philosopher scenario is set up in this way as a cosmic Doomsday Argument scenario, the BIC-based recommendation is that all those observers who live before (after) t and are aware of that should align their rational credences with the non-anthropic priors. Only those who are in the very strange epistemic situation that they are unaware of whether they live before or after t and have no reason to find it any more likely that they are any one of the observers than any other should prefer the theory T_2, which predicts a larger total number of observers, as if based on the SIA (or some variation of it). Thus, with respect to this scenario, where T_1 and T_2 agree on the history before t and disagree on it afterward, BIC-based reasoning agrees with the presumptuous philosopher's recommendation of the theory T_2 that predicts more observers as long as the presumptuous philosopher lacks crucial self-locating information (namely, as long as she does not know that she lives before t). Once she obtains this information, her rational credences simply agree with the non-anthropic input probabilities.

To conclude, rejecting the Doomsday Argument using BIC-based reasoning does not commit one to unqualified endorsement of the SIA. The implausible recommendations by the presumptuous philosophers can be avoided – in different ways for different Presumptuous Philosopher–type scenarios.

9.2.4 A First Look at the Everett Interpretation

The Everett interpretation (EI) of quantum theory can be construed as a multiverse-type theory. In Section 10.1, this interpretation is discussed more comprehensively. Here, I quickly address and rebut a challenge against BIC-based reasoning that may seem to arise from the EI – namely, that, unacceptably such reasoning entails that the EI is trivially confirmed by arbitrary empirical data.

Versions of the standard stochastic interpretation of quantum theory (SI) teach that, in each case where a measurement is performed, exactly one of the possible outcomes is realized. The probabilities for the various possible outcomes are given by their *Born weights*, which in turn can be calculated using the formalism of quantum theory. In contrast to the SI, the EI teaches that, in each case where a measurement is performed, *all* possible outcomes of the measurement are realized, but in different *branches* of a vast Everettian "multiverse," each branch associated with its distinctive Born weight. The motivation for the Everett interpretation and its implementation in the quantum theoretical formalism are discussed in Section 10.1.

Since the EI is designed to reproduce the empirical predictions of the SI in all but very special experimental circumstances that are extremely hard to bring about, one would expect that evidence that confirms (or disconfirms) the SI typically also confirms (or disconfirms) the EI to the same degree. As will be discussed in Section 10.1, there are various issues about how empirical confirmation is accommodated in the EI, and here I focus only on the most dramatic worry – namely, that, since the EI predicts that *all* possible outcome really occur, any arbitrary outcome confirms it, in contrast with the SI, which makes very specific predictions.

This would be an extremely odd result, which should make one concerned not so much about the EI but about the suggested inferences that lead to it. If it were true, it would mean that, as noted by Bradley, "[t]he Ancients could have worked out that they have overwhelming evidence for [EI] merely by realizing it was a logical possibility and observing the weather" [Bradley, 2012, p. 159]. Clearly, the case for the EI cannot possibly be as simple as that, and the reasoning that leads to trivial confirmation of the EI must be somehow fallacious.

However, as pointed out in Lewis [2007], while the naive confirmation view of the EI is obviously inadequate, reasoning along the lines of Dieks's response to the Doomsday Argument (which, as we will see in the following section, leads to the "thirder" view on the Sleeping Beauty problem), may seem to commit one to it by entailing commitment to Eq. (9.5).[1] Put as simply as possible, the argument for this claim is this: since there are multiple branches of reality according to the Everett interpretation, with many of them populated by observers, there are plausibly much more observers – or, if we choose a more fine-grained reference class of observer-stages, more observer-stages according to the EI than according to the SI.

Provided that the input probabilities $Pr(EI)$ and $Pr(SI)$ are of similar size, Eq. (9.5) then entails that we should prefer the EI over the SI, just on grounds that it entails more observer-moments. Fortunately for the Everettian, reasoning based

---

[1] Bradley endorses Peter Lewis's reasoning together with its conclusion, arguing that it is to be welcomed by the Everettian [Bradley, 2012, section 2.1]. Wilson disagrees with Bradley's analysis and argues that Everettians can escape from drawing Bradley's conclusions by relying a principled distinction between chancy and ignorance-based input probabilities [Wilson, 2014]. This is an interesting proposal that could possibly be linked fruitfully to the present BIC-based response. See, however, Bradley [2015] for criticism that Wilson's proposal would need to overcome.

on the BIC provides a rationale for not drawing this unattractive conclusion while allowing one to reject the Doomsday Argument.

To apply BIC-based reasoning to the SI/EI-dispute in a simple example, consider the stylized measurement setup considered by Lewis [2007] and Bradley [2012]: a spin-1/2 particle prepared in a state where the Born weights for the possible measurement results $+1/2$ and $-1/2$ of, say, spin in z-direction, are identical; i.e., $BW(+1/2) = BW(-1/2) = 1/2$. One can formulate the situation as a self-location problem where four possible observer-moments are to be considered – namely, EI_+, EI_-, SI_+ and SI_- – which can be read as "I am in the Everett branch where the outcome is $+1/2$," "I am in the Everett branch where the outcome is $-1/2$," "The SI holds and the outcome is $+1/2$," and "The SI holds and the outcome is $-1/2$," respectively.

Two epistemic situations are to be considered: one before the measurement outcome has been taken in, one afterward. Let us assume that, before the outcome is observed, based on our background knowledge B and certain systematic and interpretive reasons we assign an input probability $Pr(EI)$ to the Everett interpretation:

$$P(EI|EI \vee SI, B) = P(EI|EI_+ \vee EI_- \vee SI_+ \vee SI_-, B) = Pr(EI). \quad (9.6)$$

Now assume that the outcome of measurement of spin in z-direction is registered as, say, $+1/2$. A crucial assumption made by Everettians, which is relevant at this stage, is that the rational credences of being in the up- or down-branch are indeed given by the Born weights of the outcomes $+1/2$ and $-1/2$. In other words, Everettians assume that the rational credences are $P(EI_+|EI) = BW(+1/2)$ and $P(EI_-|EI) = BW(-1/2)$, where $BW(+1/2)$ and $BW(-1/2)$ are the corresponding Born weights. The challenge of justifying this assumption is known as the *probability problem* of the EI, and in the following chapter, I will argue that extant suggested solutions to this problem are unsuccessful.

For the purposes of the present discussion, we may simply grant this assumption to the Everettian. Under these circumstances, the posterior degree of belief in the EI, after having registered the outcome $+1/2$, is simply

$$P(EI|EI_+ \vee SI_+) = \frac{P(EI_+ \vee SI_+|EI) \cdot P(EI)}{P(EI_+ \vee SI_+)}$$
$$= \frac{BW(+1/2) \cdot Pr(EI)}{BW(+1/2)}$$
$$= Pr(EI). \quad (9.7)$$

Here Bayes's theorem has been used in the first line, and it has been assumed that no further interpretations besides SI and EI must be considered: $P(EI) = P(EI|EI \vee SI)$. The result Eq. (9.7) is reassuring. It means that, based on the BIC, evidence that the outcome is $+1/2$ (or $-1/2$) does *not* lead to automatic confirmation of the EI over the SI but is neutral between the two, exactly as it intuitively should, given that the EI and the SI are different interpretations of the same physical theory. The conclusion does not seem to depend on any specific features of the chosen example.

Considerations on the EI, thus, do not seem to indicate any principled difficulties with the BIC and the BIC-based rejection of the Doomsday Argument. Nor do they point to any general problematic features of the style of Bayesian reasoning used in problems of self-locating belief that is put to use in the attempts to make conceret multiverse theories testable, which were discussed in the previous two chapters.

9.3 Lessons for the Sleeping Beauty Problem

9.3.1 The Sleeping Beauty Problem and Its Suggested Solutions

Dieks [2007] discusses his rejection of the Doomsday Argument continuously with a defense of the so-called thirder response to the Sleeping Beauty problem. Since that problem is so famous and fascinating, it is worthwhile to consider the lessons of the previous sections with respect to it. As we will see, one can easily construct a version of it in which Doomsday Argument–style reasoning leads to apparent anomalous powers.

The Sleeping Beauty problem, originally due to Zuboff [1990], was made famous by Elga in the following version:

Some researchers are going to put you to sleep. During the two days that your sleep will last, they will briefly wake you up either once or twice, depending on the toss of a fair coin (Heads: once; Tails: twice). After each waking, they will put you to [sic] back to sleep with a drug that makes you forget that waking. When you are first awakened, to what degree ought you believe that the outcome of the coin toss is Heads? [Elga, 2000, p. 143]

Opinions are split over the correct answer. The two candidate rational credences for Beauty ("you," in Elga's example) with respect to *Heads* are $1/2$ and $1/3$, both of which have substantial support in the literature. It is impossible to do justice to that by now very extensive literature; see Titelbaum [2013b] for a condensed overview. For some defenses of the $1/3$ view, see Dorr [2002], Horgan [2004], Hitchcock [2004], Draper and Pust [2008], Titelbaum [2008], Schulz [2010], and Titelbaum [2013a]. For some criticisms of it, sometimes combined with an endorsement of the $1/2$ view, see Jenkins [2005], White [2000], Bradley and Leitgeb [2006], Bradley [2011], Bradley [2012]. Further important studies include Kierland and Monton [2005]; Briggs [2010]; Ross [2010], and Schwarz [2015].

The simplest arguments in favor of the $1/3$ view, both proposed and endorsed by Elga [2000], are the following (where, in accordance with convention, Beauty's first awakening is supposed to take place on Monday and the second, which occurs only if the coin falls *Tails*, on Tuesday): first, if the experiment is repeated many times, approximately $1/3$ of the awakenings are *Heads* awakenings; second, on the $1/2$ view championed by Lewis [2001] and Bradley [2012], if on Monday someone tells Beauty it is Monday, standard Bayesian conditioning tells her to shift her credence with respect to *Heads* from $1/2$ to $2/3$ (I use "P^-" to denote Beauty's credences on Monday before she knows it is Monday and "P^+" to denote her credences when she does know it is Monday):

$$P^-(Heads) = 1/2 \xrightarrow{Monday} P^+(Heads) \tag{9.8}$$
$$= P^-(Heads|Monday) = 2/3.$$

This means that Beauty's rational credence with respect to *Heads* differs from its objective chance $Pr(Heads) = 1/2$, even though, knowing it is Monday, Beauty is now fully oriented about her temporal position, in apparent contradiction with David Lewis's famous Principal Principle [Lewis, 1986a]. The essential statement of Lewis's response in Lewis [2001], defended by Bradley [2011], is that Beauty acquires inadmissible evidence when she learns that it is Monday, which disqualifies straightforward use of the Principal Principle

The most important argument against the $1/3$ view is that, in analogy with Dieks's response to the Doomsday Argument, it entails commitment to Eq. (9.5) as adapted to the Sleeping Beauty problem, which, when used as a general principle, runs into the Presumptuous Philosophy problem. The essential analogy between the thirder position on Sleeping Beauty and Dieks's response to the Doomsday Argument is that both identify the input probabilities – in Beauty's case, the chances $Pr(Heads)$ and $Pr(Tails)$; in the Doomsday case, the probabilities $Pr(H_1)$ and $Pr(H_2)$ – with the credences one should have *when*, not before, one has the relevant bits of self-locating information "it is Monday" and "*n* is my birthrank."

Conversely, the halfer position on Sleeping Beauty and the Doomsday Argument identify the input probabilities with the credences one should have *before*, not when, one has the self-locating information at issue. Both Dieks [2007] and Bradley [2012] highlight these connections between viable positions on Sleeping Beauty and the Doomsday Argument – Dieks while defending the thirder position and rejecting the Doomsday Argument, Bradley while defending the Lewisian halfer position and endorsing the Doomsday Argument.

In addition to the thirder position, as defended by Elga and Dieks, and the halfer position, as defended by Lewis and Bradley, there is an additional, third, position on Sleeping Beauty. Its adherents [Halpern, 2005; Bostrom, 2007; Meacham, 2008; Cozic, 2011] concur with Lewisian halfers that Beauty's credence with respect to *Heads* should be $1/2$ when she awakes, but they also claim that it should *remain* $1/2$ when she learns that it is Monday. Double-halfing is an interesting alternative to thirding and Lewisian halfing. It parallels those responses to the Doomsday Argument according to which learning one's birthrank should not change one's credences with respect to the competing hypotheses H_1 and H_2.

Unfortunately, double-halfing faces the following serious problem [Bradley, 2011; Titelbaum, 2012]: if a coin is tossed on Tuesday evening in addition to the first one (tossed on Sunday or Monday evening), then, according to the double-halfer position, Beauty's credence with respect to "Today's coin will fall *Heads*" when awakening must be larger than $1/2$: on the double-halfer position, if $Heads_1/Tails_1$ is the outcome of the first toss,

$$P(Monday, Heads_1) = 1/2$$

and

$$P(Monday, Heads_1) + P(Tuesday, Heads_1) = 1/2.$$

If self-locating indifference is assumed, this becomes

$$P(Monday, Heads_1) = P(Tuesday, Heads_1) = 1/4.$$

The supposed second coin toss would occur on Tuesday. If we denote its outcome by $Heads_2/Tails_2$ and assume, as we unproblematically can, that this outcome is uncorrelated with whether Beauty is woken once or twice, we obtain

$$P(Tuesday, Heads_2) = P(Tuesday, Tails_2) = 1/8,$$

where the last equality obtains if indifference is assumed. It follows that Beauty's credence in "Today's coin will fall *Heads*" must be

$$P(Monday, Heads_1) + P(Tuesday, Heads_2) = 5/8$$

if indifference is assumed, and, in any case, larger than $1/2$.

This is an unacceptable result because, when Beauty learns what day it is, her rational credence with respect to this proposition drops to $1/2$, no matter what she learns. Thus, according to the double-halfer position, Beauty's epistemic stance with respect to "Today's coin will fall *Heads*" is oddly unstable before she is informed what day it is. For a further strong criticism of double-halfing, see Conitzer [2015a].

Having ruled out the double-halfer position, what is the most attractive remaining view of the Sleeping Beauty problem from the perspective of the BIC? This question translates into whether objective chances (like those in a fair die toss) translate into the rational credences of those whose background knowledge provides full self-locating information or whether, as argued by Bradley [2011], self-locating information can be, in the language of the Principal Principle, "inadmissible," which would mean that objective chances sometimes prescribe the rational credences only in situations where agents *lack* some bit of self-locating information.

To answer this question, it is useful to recall the discussion of the Doomsday Argument in the first section of this chapter. There we identified credences that are as if someone had anomalous causal powers as reflecting badly on the type of reasoning that led to them. This turns out to be a useful strategy in the context of the Sleeping Beauty problem as well.

9.3.2 Sleeping Beauty with Anomalous Causal Powers

Given the similarities between Lewisian halfing and the Doomsday Argument, can we construct a version of the Sleeping Beauty problem in which Lewisian halfing recommends credences that are as if someone (say, Beauty herself) had anomalous causal powers? We can, as the *Choosing Beauty* problem shows:

Choosing Beauty (CB): As in the original Sleeping Beauty problem, Beauty is woken either once (on Monday) or twice (on Monday and Tuesday), depending on the outcome of a fair coin toss (one awakening if the coin comes up *Heads*, two if it comes up *Tails*). All her memories of any previous awakenings during the trial are erased by a drug whenever she is put to sleep. This time, however, *two* coin tosses are performed, both on Monday evening.

After having been woken on Monday, Beauty is told that it is Monday and is asked to choose whether the outcome of the first or the second coin toss to be performed the same evening is to count as relevant for whether or not she is woken on Tuesday. In accordance with the outcome of that coin toss, she is woken or not woken on Tuesday.

Let us refer to Beauty's two possible choices as C_1 ("The first coin toss counts") and C_2 ("The second coin toss counts"). Now consider one of the coin tosses – say, the first – and consider Beauty's rational credences with respect to its possible outcomes $Heads_1$ and $Tails_1$, as assessed from the point of view of Lewisian halfing. According to this position as applied in the original SB problem, Beauty's rational credence with respect to the outcome *Heads* on the chosen coin is $1/2$ on Monday morning and $2/3$ after she has been told that it is Monday.

There is no reason to suppose that her rational credences about the possible outcomes of the coin toss she does *not* choose are at any stage different from $1/2$. To conclude, by the standards of Lewisian halfing, after Beauty has been told that it is Monday, her rational conditional credences with respect to $Heads_1$ are $P^+(Heads_1|C_1) = \frac{2}{3}$ and $P^+(Heads_1|C_2) = \frac{1}{2}$, and similarly (mutatis mutandis) for the other possible outcomes of the two tosses.

What seems odd about these credences is not only that there is some *fixed* future (or past, if the coins are tossed on Sunday) coin toss with respect to which, as in the original SB puzzle, $P^+(Heads) = 2/3 \neq Pr(Heads) = 1/2$ but also that the identity of this coin toss (which one it is) depends on a choice Beauty makes at the very same stage. The thirder position concurs with Lewisian halfing that there is *some* stage at which Beauty's credence with respect *Heads* for the chosen coin should depart from $1/2$ in that, according to thirdism, $P^-(Heads_1|C_1) = 1/3$ and $P^-(Heads_2|C_2) = 1/3$ for her credences before she is told it is Monday.

However, this is not a situation where Beauty can be given the choice between C_1 and C_2, for giving her the choice means telling her it is Monday. On Tuesday, the coin toss whose outcome decides whether she is woken once or twice has already been tossed (and, if she has been woken, fallen *Tails*). So, giving her the choice between C_1 and C_2 tells her it is Monday and, thereby, lets her credence shift to $P^+(Heads_1) = P^-(Heads_1|C_1, Monday) = P^-(Heads_1|C_2, Monday) = 1/2$. Accordingly, as soon as Beauty *can* make her choice, her rational credences about possible outcomes are all equal to $1/2$, so that, unlike according to Lewisian halfing, there is no stage at which she simultaneously has the choice between C_1 and C_2 and a rational credence with respect to *Heads* that differs from $1/2$ for some toss.

In order to make the weird and counterintuitive consequences of Lewisian halfing as applied to the CB problem most vivid, it is useful to have in mind the "extreme" version of CB ("Extreme Sleeping Beauty," [Bostrom, 2007, p. 66], where, if the chosen coin comes up *Tails*, Beauty is woken not twice but N times on subsequent days for some very large number $N \gg 1$. In that scenario, according to Lewisian halfing, Beauty's rational conditional credences when she learns it is Monday are $P^+(Heads_1|C_1) = \frac{N-1}{N} \approx 1$ and $P^+(Tails_1|C_1) = \frac{1}{N} \approx 0$ and, trivially, $P^+(Heads_1|C_2) = P^+(Tails_1|C_2) = 1/2$ (and equivalently under exchange of the indices 1 and 2).

What makes Beauty's credences odd here is that they are analogous to those of a person who is in a position to choose between two coins to be tossed as to which of them should be *manipulated* (by affecting its internal mass distribution, say), such that the outcome of its toss becomes almost certainly *Heads*. By way of manipulating the coin, such a person would be able to *causally influence* the outcome of its toss, and the parallel between that person's rational credences and Beauty's according to Lewisian halfing confirm that the latter are indeed as if Beauty had anomalous causal powers in the sense discussed. In particular, just as it would be rational for that person to manipulate the first coin to be tossed by modifying its internal mass distribution if she wanted its outcome to be *Heads*, according to Lewisian halfing, it would apparently be rational for Beauty to make the choice C_1 if she wished that the outcome of the first coin toss should be *Heads*$_1$. Accordingly, yet implausibly, putting oneself in the same situation as Beauty and choosing the first coin would be practically equivalent to manipulating it directly.

9.3.3 Causal Decision Theory to the Rescue?

The only available strategy for proponents of Lewisian halfing to avoid this unattractive recommendation might be to appeal to causal decision theory. More specifically, they might suggest that even though Beauty's rational credences are as if she had anomalous causal powers, these odd credences are not the ones that should actually guide her actions and decisions.

To argue for this, perhaps seemingly rather odd, position, Lewisian halfers might compare Beauty's situation in CB to that of a subject in a medical Newcomb problem. In a typical medical Newcomb problem, there is some disease B for which bodily feature A is a symptom, such that A's first appearance reliably indicates that the person will fall ill with B some days later. Given the available statistical data, A and B are positively probabilistically correlated in that $Pr(B|A) > Pr(B)$, which manifests itself in the subject's rational credences $P(B|A) > P(B)$. This correlation, however, is not due to A's *causing* B, but, instead, due to there being some bodily state C – the presence of certain bacteria in the organism, say – that typically leads to both A and B and that *screens off* A from B in that $Pr(B|A, C) = Pr(B|C)$.

Before looking into this proposal a bit more thoroughly, it is worthwhile to note that it does not seem very promising in light of the literature on the Sleeping Beauty problem and decision theory. Briggs [2010], building on earlier work by Hitchcock [2004], shows that the Lewisian halfer position in combination with causal decision theory leads to a diachronic Dutch book – i.e., a sure loss in some cleverly chosen gambling setup, usually interpreted as a signal of irrationality.

However, Briggs [2010] also shows that the thirder position in combination with *evidential decision theory*, the main alternative to causal decision theory, leads to a diachronic Dutch book as well. Since there are arguments for preferring evidential decision theory over causal decision theory (and of course, vice versa [Pearl, 2000]), one may also resort to decision-theoretic considerations in order to *attack*, rather than defend, the thirder position. The discussion is further complicated by an argument due to Conitzer [2015b], according to which evidential decision theory leads to irrational behavior in the Sleeping Beauty problem, independently of whether one accepts the halfer or thirder position.

Let us return to the question whether the Lewisian halfer may resort to causal decision theory to evade the counterintuitive recommendations of her position with respect to the Choosing Beauty problem. The motivation for preferring causal decision theory over evidential decision theory, recall, is that it more straightforwardly delivers the correct reactions to problems with the same causal structure as a medical Newcomb problem. Evidently, for a subject that faces a medical Newcomb problem, taking precautions against A that are not effective against C is an ineffective strategy for avoiding B. It is controversial in the philosophical literature whether evidential decision theory, properly interpreted and applied, can reproduce the correct recommendations for rational action in such problems or whether causal decision theory is needed as an alternative with a fundamentally different starting point.[2]

The debate between evidential and causal decision theory aside, there is nothing particularly mysterious about medial Newcomb problems: it is unsurprising that they very frequently arise in practice, and it is uncontroversial that the rational course of action in them is not to combat the symptom A, at least not without also fighting the cause C.

The crucial parallels that Lewisian halfers might point out between Choosing Beauty and the medical Newcomb problem are first, that an agent can supposedly control some variable – the presence of the symptom A in the medical Newcomb problem and the outcome of the choice between C_1 and C_2 in CB; second, that the value of that variable is probabilistically correlated with some later event – the disease B in the medical Newcomb problem and the outcome of the coin toss chosen by Beauty in CB; and, third, that there is no causal influence from the controllable variable to the later event.

Pointing out these parallels, proponents of Lewisian halfing might argue that Beauty in the CB scenario should regard the outcome of her choice between C_1 and C_2 as merely *symptomatic* of the outcome of the chosen coin toss, just as the subject in a medical Newcomb problem should regard the symptom A as symptomatic of whether she or he will fall ill with B some days later. And indeed, this seems to be Bostrom's perspective on Lazy Adam, with respect to which he recommends that Adam may regard his "choice [as] an *indication* of a coincidence" [Bostrom, 2001, p. 371], namely, one between the outcome of the choice itself and the later course of events. So, given all these parallels, is the CB scenario as seen from the perspective of Lewisian halfing perhaps no more odd and problematic than a medical Newcomb problem?

Arguably not, for at some point, the parallels end. In a medical Newcomb problem, correlations are non-mysterious and rational actions uncontroversial due to there being the state C that, as explained, *screens off* A from B in that $Pr(B|A,C) = Pr(B|C)$. If medical research finds no state C with the required properties, the conditions for a medical Newcomb problem are not met, and taking precautions against A is considered a potentially effective means for preventing B.

In the CB scenario, Lewisian halfers cannot point to any state or event C such that $P^+(Heads_1|C_1,C) = P^+(Heads_1|C_1,C)$; i.e., there is just no reason to expect screening off between C_1 and $Heads_1$ as far as Beauty's rational credences

---

[2] See Lewis [1981] and Price [1996] for examples of important contributions on the two different sides of the debate.

are concerned. So, unlike a subject in a medical Newcomb problem, if Beauty accepts the Lewisian halfer's recommendations, she has no comparable reasons to not take the probabilities $P^+(Heads_1|C_1)$ and $P^+(Heads_1|C_2)$ as the ones to base her rational actions on. This suggests that, according to Lewisian halfing as applied to the CB scenario, not only Beauty's rational credences but also her rational actions are as if she had anomalous causal powers.

The implausibility of these consequences of Lewisian halfing when applied to CB suggest that, from the point of view of the BIC, Beauty's self-locating information about which day of the week it is counts as "admissible." The objective chances of the two possible outcomes of the die toss translate into her rational credences concerning those outcomes in her epistemic situation where she has complete self-locating information, not when she lacks some. This allows us to use BIC-based reasoning as the key to simultaneously reject the Doomsday Argument, reject the presumptuous philosopher's reasoning, reject the view that the Everett interpretation of quantum theory is automatically confirmed by arbitrary evidence, and endorse the thirder position in the Sleeping Beauty problem. Arguably, all of these recommendations are plausible.

Part IV

Wider Still and Wilder

10

Other Multiverses

Having discussed the prospects for multiverse theories that pertain to Tegmark's "level II" in previous chapter, I now turn to his level III and level IV multiverses in this chapter. Section 10.1 investigates the level III multiverse entailed by the Everett interpretation of quantum theory; Section 10.3 discusses the level IV mathematical multiverse. In between these two sections, I address yet another multiverse proposal in Section 10.2, which is not very often labeled as such – namely, philosopher David Lewis's multiverse of all possible worlds, as entailed by his peculiar metaphysics of modality called *modal realism*. The recurring thread that connects these discussions with the previous two chapters is the continuing importance of matters of self-locating belief. The recurring outcome of the present discussion of those multiverse proposals is that, to the extent that they are coherent at all, they are still not coherently believable when taken seriously.

10.1 The Everettian Quantum Multiverse

10.1.1 The Quantum Enigma

Quantum theory is the theoretical framework in which we currently formulate theories of fundamental physics. The electroweak theory and quantum chromodynamics, the two pillars of the Standard Model of elementary particle physics, are quantum theories. The fact that Einstein's general theory of relativity is *not* a quantum theory and is not as straightforwardly "quantized" as one might hope is one of the most pressing problems of contemporary fundamental physics. Physicists, in general, expect future theories of fundamental physics to be quantum theories as well. String theory, for instance, is a quantum theory of strings and branes.

Yet, despite quantum theory's overwhelming predictive and explanatory success, the question of why this theoretical framework works as well as it manifestly does remains mysterious. The objects in a quantum theory are characterized by *dynamical variables* – for example, for a particle like an electron, its position, momentum, and spin. But quantum theory does not assign values to these variables, nor does it specify their dynamics. What it does help us assign are so-called quantum states ψ, which ascribe probabilities to the different possible values that dynamical variables may have. There tends to be fantastic agreement between these probabilities and the relative frequencies obtained by repeated measurements of the dynamical variables in question.

The most straightforward interpretation of these probabilities is that they express incomplete information about the actual, underlying values of the dynamical variables. Unfortunately, this straightforward interpretation is fraught with difficulties. If one tries to express a quantum state as a probability distribution in phase space (i.e., position-momentum space), one obtains the so-called *Wigner function*. This function is not a real probability distribution – it is called a *quasi-probability distribution* – because it is usually negative in some regions of phase space, which means that it cannot be interpreted straightforwardly as expressing limited information about the system's true location in phase space. In practice, one considers only the probabilities assigned to the possible values of those variables that are *measured* in a given experiment, not those assigned to arbitrary variables. In this role, quantum theory enjoys unparalleled success.

This success, however, is extremely puzzling: how can the world possibly be such that the quantum theoretical "algorithm" to predict and account for measurement outcomes works as well as it manifestly does. Answering this question is particularly difficult in the light of various *no-go* theorems that show that, given various rather plausible constraints, there cannot be any possible variable configuration that would account for the empirical success of quantum theory.

Notably, any such configuration has to be *nonlocal* in the sense that correlations between distant events cannot be explained by either mutual causal influences or common causes in the joint past [Bell, 1964; Clauser et al., 1969; Greenberger et al., 1989]; and it has to be *contextual* in the sense that the probability of a certain value would depend on the choice of variables that are simultaneously measured [Bell, 1966; Kochen and Specker, 1967; Spekkens, 2005]. Section 9.2.3 featured the "stochastic interpretation" SI as the standard (one-world) view of quantum theory. But the results just mentioned indicate that it is very difficult to specify what this interpretation might concretely look like in detail if it is to be more than the standard quantum algorithm the success of which is mysterious.

Historically, the most influential view on quantum theory consisted in simply refusing to give an account of how the world could possibly be such that the success of quantum theory becomes understandable. This perspective is part of the famous *Copenhagen interpretation*, a view that is associated mostly with Niels Bohr and to some extent Werner Heisenberg, but extremely difficult to actually pin down. It lives on in the radically subjectivist *quantum Bayesian* point of view [Fuchs and Schack, 2015] and in the nuanced pragmatist interpretation recently developed by Richard Healey [2017]. The "therapeutic" perspective that I myself tentatively advocated in my book [Friederich, 2014] is also a variety of this view.

It is not really clear, though, to what degree refusing to provide an account of what the world might fundamentally be like such that the success of quantum theory becomes understandable is a coherent option at all, unless, say, one is also prepared to sacrifice everyday common-sense realism.

There are well-developed accounts of what the world *might* be like such that the success of quantum theory would become understandable. The most-discussed ones are Bohmian mechanics (also known as pilot wave theory) [Bohm, 1952] and the Ghirardi-Rimini-Weber (GRW) spontaneous localization model [Ghirardi et al., 1986]. Both of them incorporate nonlocality and contextuality, and both, notably Bohmian mechanics, have avid supporters. Both, however, also *add* additional elements to the quantum theoretical formalism, which make them less attractive to many physicists. In the case of the GRW model, these additional elements are relatively ad hoc dynamics for wave function collapse, which, for example, lead to a violation of energy conservation. In the case of pilot wave theory, they are somewhat difficult to generalize to the relativistic setting, where there is no universally preferred time parameter.

In view of the difficulties facing nonrealist approaches to quantum theory on the one hand and realist approaches like pilot wave theory and the GRW model on the other, the Everett interpretation [Everett, 1957] (EI; the most-discussed modern version is the one defended by Wallace [2012]) promises to combine all the advantages that the no-go theorems seemed to exclude: a foundational account of quantum theory that is realist and does not require any amendments to or adjustments of the theoretical formalism.[1] But there is a price to be paid for this elegant solution to the foundational problems of quantum theory: one must postulate the existence of many other, distinct "branches" or reality ("many worlds") and thereby assume that we live in a type of multiverse whose potential existence we did not anticipate prior to the advent of quantum theory.

10.1.2 Defining Everettian Branches

The EI entails a multiverse not in the sense of a spatiotemporally disjoint or distant subuniverses as the cosmological multiverse discussed in the previous chapters.

---

[1] It should be noted that there are variants of the Everett interpretation that do make amendments or adjustments to the formalism, notably the "many-minds" view considered by Albert and Loewer [1988].

Rather, it entails a multiverse in the sense of a system of distinct "branches" of the universal wave function $|\Psi\rangle$, which it regards as providing a complete description of the world. The nature of these branches can be illustrated by considering a situation where some typical quantum micro-system (e.g., an electron) interacts with its environment in a characteristic way. Importantly, that environment can be either an external environment (for a micro-system) or the internal environment (the internal degrees of freedom for a macro-system). Generically, a phenomenon called *decoherence* occurs in that situation, which means that for some specific orthonormal basis $\{|\psi_i\rangle\}$ of system states, the interaction dynamics have the form

$$|\psi_i\rangle \otimes |\phi\rangle \mapsto |\psi_i\rangle \otimes |\phi_i\rangle, \tag{10.1}$$

where

$$\langle\phi_j|\phi_i\rangle \approx \delta_{i,j}. \tag{10.2}$$

The dynamics of Eq. (10.1) unfold much faster than the individual dynamics of the system and its environment, respectively.

After the interaction, the states of the system and its environment are entangled, which means that the system is no longer described by an individual pure state at all. One can, however, ascribe the reduced density matrix to it, obtained by performing the trace over the environment degrees of freedom. For a system setting out in an arbitrary state $|\psi\rangle$, that density matrix will now be approximately diagonal in the basis $\{|\psi_i\rangle\}$; i.e., it will have the approximate form

$$\rho = |\psi\rangle\langle\psi| = \sum_i |c_i|^2 |\psi_i\rangle\langle\psi_i|, \tag{10.3}$$

with only comparatively small corrections from mixed terms $|\psi_i\rangle\langle\psi_j|$ with $i \neq j$. The almost complete absence of these terms means that there are essentially no effects from *interference* by future superposition of the distinct states $|\psi_i\rangle$. The factors $|c_i|^2$ are the *Born weights* mentioned in Section 9.2.3. As mentioned there, in a measurement setting, their physical role is to function as probabilities of the possible measurement outcomes.

The permanent action of decoherence, which suppresses interference effects, means that the *universal* quantum state $|\Psi\rangle$ will experience *branching*. Mathematically, branching can be spelled out as follows: For some family of disjoint projection operators $\hat\Pi_i$ which sum to the identity, one considers their associated (Heisenberg picture) time-dependent projection operators $\hat\Pi_i(j)$, where j is a suitably discretized time variable. Wallace defines as the *branching criterion* the statement that if

$$\hat\Pi_{i'}(j')\hat\Pi_i(j_1)|\Psi\rangle \tag{10.4}$$

and

$$\hat{\Pi}_{i'}(j')\hat{\Pi}_i(j_2)|\Psi\rangle \tag{10.5}$$

are both non-zero, then $j_1 = j_2$ [Wallace, 2012, p. 92]. A time-ordered sequence of projection operators $\hat{\Pi}_i(j)$ is called a *history*. A history counts as *realized* in the Everettian multiverse if the expectation values of the associated sequence of projection operators are non-zero. That the branching criterion obtains because of decoherence means that if two realized histories coincide at a certain time j, they coincide at all earlier times $j' < j$, but not necessarily at later times $j' > j$. In that sense, the branching criterion, indeed, entails the existence of a "branching" structure of the universal quantum state Ψ.

The core idea of the modern version of the Everett interpretation is that the observations about decoherence just sketched, combined with a realist view of the universal quantum state $|\Psi\rangle$, suffice to provide a compelling picture of physical reality that demystifies the empirical success of quantum theory. *All* realized histories are physically real according to the Everettian, in the sense that they are an emergent feature of the universal quantum state that arises due to decoherence. The appearance of an approximately classical, single world is simply due to the fact that our experience is restricted to our own "branch" of the universal wave function.

Traditionally, one distinguishes between two main challenges to the Everett interpretation: the first, nowadays called the "ontological problem" (and traditionally the "preferred basis problem"), is the one just addressed: to identify "branches" within the universal quantum state that behave as quasi-classical "worlds." Contemporary Everettians seem to be more or less unanimous in regarding (some version of) the appeal to decoherence just sketched as the solution to this problem. The second challenge is the so-called probability problem: accounting for how quantum probabilities arise in the Everett interpretation and, specifically, why the Born weights $|c_i|^2$ associated with different branches i at a given time would effectively function as probabilities for an Everettian agent. Meeting this challenge is necessary to ensure that the Everett interpretation can qualify as empirically adequate.

In the following section, I will argue that extant proposals to solve the probability problem fail. But before addressing those proposals, it is useful to realize that, in the absence of a compelling solution to the probability problem, the ontological problem cannot really be regarded as solved either; the reason for this is that decoherence and (approximate) branching occur only inasmuch as there is a standard by which one can regard the non-diagonal "interference" terms in the reduced density matrix of the system undergoing decoherence, which are proportional to $|\psi_i\rangle\langle\psi_j|$ with $i \neq j$, as "small" and effectively negligible from the physical point of view.

However, as long as no physical interpretation of the expansion coefficients c_i of the universal wave function in some basis has been given, it is not clear by what standard something can count as so small as to not really physically matter. The correct physical interpretation of the coefficients, as we know, is that their modulus squared, $|c_i|^2$, has the physical role of functioning as a probability. Thus, it seems that in the absence of a compelling solution to the probability problem – the problem of specifying how the $|c_i|^2$ come to play their roles as probabilities – there cannot be a compelling solution to the ontological problem.

Worse, extant attempts to solve the probability problem tend to *presuppose* the existence of a suitable branching structure; i.e., they assume that a successful solution to the ontological problem has already been provided. Zurek [2005, section 7.B], Baker [2007], and Kent [2010] point out that this gives rise to a circularity in the version of the EI that has been discussed most in recent years: the Deutsch-Wallace decision-theoretic approach [Deutsch, 1999; Wallace, 2007, 2012], which will be briefly discussed in the next section. Even more problematically, as pointed out by Dawid and Thébault [2015], some suggested solutions to the probability problem, again including the Deutsch-Wallace approach, apply only to the coefficients c_i inasmuch as those give rise to non-negligible branch weights $|c_i|^2$. They do not provide any physical interpretations of the non-diagonal "interference" terms proportional to $|\psi_i\rangle\langle\psi_j|$ with $i \neq j$ and, a fortiori, do not provide any license to regard those as of minor physical importance. This, however, is a required condition for the ontological problem to be solved. Therefore, as Dawid and Thébault argue, attempts to solve the probability problem by presupposing branching structure are not merely circular but even incoherent.

In what follows, we look at the probability problem in the EI through the lens of the considerations on multiverse theories testing and self-locating belief developed in earlier chapters of this book. As I shall argue, in the light of those considerations, extant attempts to solve the probability problem in EI are not compelling.

10.1.3 Attempts to Solve the Probability Problem

One can distinguish between three types of attempts to solve the probability problem of Everettian quantum theory: first, as Lev Vaidman does in his earlier writing [Vaidman, 1998], one can simply *postulate* as a primitive fact of rationality that the rational self-locating credences about being in specific branches of the Everettian multiverse are given by the respective Born weights of the branches; second, the decision-theoretic approach pioneered by Deutsch [1999] and refined by Saunders [2004] and Wallace [2007] argues that the rational decisions of agents in an Everettian world will make use of the Born weights of branches as probabilities; and third, as recently suggested by Sebens and Carroll [2018], on the one hand,

and McQueen and Vaidman [2019], on the other, one can try to derive the conclusion that a rational epistemic agent's self-locating credences in the Everettian multiverse conform to the Born weights from more basic principles of self-locating belief that have a higher degree of intuitive plausibility.

The Born Rule as a Primitive Rationality Principle?

To motivate an approach of the first type, Vaidman introduces the term "measure of existence" [Vaidman, 1998, section 9] for the Born weight of a branch of the universal wave function and then simply stipulates the "probability postulate" according to which "If a world with a measure μ splits into several worlds then the probability ... for a sentient being to find itself in a world with measure μ_i (one of these several worlds) is equal to μ_i/μ" [Vaidman, 1998, p. 254]. The profound-sounding term "measure of existence" should not distract us from the fact that Vaidman simply *postulates* the Born rule here as a primitive rationality principle that purportedly applies to agents in an Everettian multiverse. A similar move – adding the assumption that the Born rule as a separate empirical principle to the quantum theoretical formalism interpreted along Everettian lines – is proposed by Greaves and Myrvold [2010].

I do not think this is an acceptable move. Plausibly, on a single-world interpretation of quantum theory, quantum theory is empirically adequate only if the relative frequencies of events to which it applies correspond approximately to the Born weights. And, indeed, this is the finding of an immense number of empirical tests of quantum theory, applied in different domains from high-energy physics over solid-state physics to quantum chemistry. In a *many worlds* view, in contrast, it is not a universal feature of the branches created by decoherence that the relative frequencies of the events that occur in them are close to the Born weights. Multiplying worlds beyond those in one-world interpretations of quantum theory by adopting a *many worlds* view thus raises, at least prima facie, legitimate worries concerning the continued empirical adequacy of the formalism. Those worries are aggravated when one observes that, in stylized cases where branches are defined sharply, such that precise branch numbers can be defined, verdicts derived from the principle of self-locating indifference Eq. (9.2) do not, in general, agree with those derived from quantum theory, applied properly: the observations made by *most* (or *typical*) observers across branches will not be the same as those made by observers in the branches with the largest Born weights.

Proponents of the Everett interpretation respond to this latter worry, articulated succinctly by Dizadji-Bahmani [2015], by highlighting that precise branch numbers are not well defined in more realistic applications of Everettian quantum theory (e.g., Wallace [2012, section 3.11]). There is thus no good sense in which observers in branches where relative frequencies do *not* approximately match the Born weights are more numerous than observers in branches where relative frequencies do approximately match the Born weights. But this response, again, unfairly shifts the burden of proof to those who raise the suspicion that the Everett interpretation might not be empirically adequate due to the fact that it postulates branches in which relative frequencies do not approximately match the Born weights. The onus is on the Everettian to demonstrate that we can be reasonably confident to not find ourselves in one of those branches if the EI is true.

The Decision-Theoretic Program

The decision-theoretic program of solving the probability problem in Everettian quantum theory, pioneered by David Deutsch [1999] and further developed by Simon Saunders [2004] and David Wallace [2007], accepts this challenge and tries to answer it. The core idea of this approach is that the Born rule "follows" in some sense from Everettian quantum mechanics even though the Born weights do not represent any kind of probability of existence. Rather, according to this approach, using the Born weights as self-locating probabilities is the only possible strategy on which to ground rational betting behavior that is available to an agent who endorses the EI and believes that it correctly describes reality.

To arrive at this conclusion, Deutsch, Saunders, and Wallace need to introduce fairly specific rationality conditions that determine what it takes to bet rationally. Among those conditions are *diachronic consistency* – which enforces that rational betting on the next measurement outcome must not depend on the branching structure in that measurement's future – and *branching indifference* – which dictates that agents are indifferent about whether they are subjected to branching or not, provided the post-branching rewards are the same. But these and other rules used by Deutsch, Saunders, and Wallace are controversial [Price, 1996; Kent, 2010; Dizadji-Bahmani, 2015]. For example, Sebens and Carroll [2018, appendix B] reject diachronic consistency as a rationality criterion because, as they convincingly argue, it is violated in independently attractive instances of self-locating reasoning in well-defined multiverse setups. Dizadji-Bahmani [2015] attacks branching indifference because it is incompatible with self-locating indifference Eq. (9.2) in the stylized scenarios mentioned earlier where branch numbers are precisely defined.

Another difficulty for the decision-theoretic approach is that it is doubtful whether decision-theoretic considerations by themselves can possibly be sufficient to establish the Born rule in the required sense. All that one can hope to establish with the derivation of the Born rule in the decision-theoretic program is that certain types of betting behavior are rational in an Everettian multiverse. Where some type of betting behavior is based on probabilities that are intended to roughly match certain relative frequencies as predicted by some theory, finding that betting behavior to be unsuccessful indicates that the actual relative frequencies are different from those predicted by the theory. Such a finding would therefore be interpreted as evidence that the theory is empirically inadequate.

In the decision-theoretic approach to the probability problem in the EI, however, the link between probability and relative frequency is cut, and this limits the relevance of decision-theoretic considerations with respect to the theory's empirical adequacy. In other words, if the betting behavior based on Born weights as probabilities turned out to be unsuccessful, there would – troublingly – be no reason why the Everettian would have to interpret this as evidence against quantum theory. As pointed out by Dawid and Thébault, "[n]o deductive step is available that leads from the statement that only betting on a specific statistical data distribution is rational to the statement that this distribution will in fact be found in experiments" [Dawid and Thébault, 2014, p. 57]. It seems difficult to deny that we should regard quantum theory as falsified if we repeatedly found empirical data that do not conform to predictions derived from the Born rule. But, as Dawid and Thébault conclude, the decision-theoretic approach "gives us no justification at all for withdrawing the predictions according to the Born rule in the face of data that disagrees with the Born rule" [Dawid and Thébault, 2014, p. 57], contrary to what would be good scientific practice. Along similar lines, Adlam [2014] argues that the decision-theoretic program is unsuccessful in that it fails to establish how agents in an Everettian multiverse could possibly be able to obtain any empirical confirmation of quantum theory.

Deriving the Born Rule from Considerations about Self-Locating Belief?

A strategy for motivating the Born rule that more closely mirrors the attempts to make multiverse theories empirically testable that were discussed in Chapters 7 and 8 has recently been proposed by Sebens and Carroll [2018]. Their approach parallels the decision-theoretic program in presupposing that the ontological problem can be assumed to be solved by an appeal to decoherence. Sebens and Carroll do not rely on any decision-theoretic considerations, however, but they try to derive the Born rule from the structure of the Everettian multiverse by using only one additional principle of rational self-locating belief, which they call the *epistemic separability principle* (ESP). Because their approach is relatively novel and interesting to assess from the point of view of the considerations on self-locating belief developed in Chapter 9, I devote some more space to it in what follows.

Sebens and Carroll characterize the "gist" of their principle ESP as the idea that "[t]he credence one should assign to being any one of several observers having identical experiences is independent of the state of the environment" [Sebens and Carroll, 2018, p. 40]. In other words, according to ESP, an agent's rational self-locating credences depend only on the states of those subsystems S of the universe

U that contain observers with internally qualitatively identical states as the agent's and not on the state of the rest of U. The full formulation of ESP is as follows:

ESP: Suppose that [the] universe U contains within it a set of subsystems S such that every agent in an internally qualitatively identical state to agent A is located in some subsystem which is an element of S. The probability that A ought to assign to being located in a particular subsystem $X \in S$ given that they are in U is identical in any possible universe which also contains subsystems S in the same exact states (and does not contain any copies of the agent in an internally qualitatively identical state that are not located in S). [Sebens and Carroll, 2018, p. 40]

$$P(X|U) = P(X|S) \tag{10.6}$$

Here, the conditional probability $P(X|U)$ is to be understood as the self-locating probability of being in subsystem X, given that one inhabits the universe U. Similarly, $P(X|S)$ is the self-locating probability of being in the subsystem X, given that one inhabits one of the subsystems that are elements of S, not knowing which.

The motivating idea behind ESP, similar to that behind self-locating indifference Eq. (9.2) and BIC as a principle of reference class choice, seems to be that rational self-locating credences cannot possibly depend on anything beyond the epistemic agent's empirical access. Sebens and Carroll illustrate this motivation by means of an example due to Elga [2004] that features the characters Dr. Evil, located on the moon, and his duplicate Dup, located on Earth (U is the state of the entire universe):

The probability $P(\text{I'm Dr. Evil}|U)$ should not depend on what's happening deep inside the Earth's core or what's happening on the distant planet Neptune or any other remote occurrences. Unless, of course, the actions of the Earth's core cause earthquakes which the terrestrial Dup feels but the lunar Dr. Evil does not. Or, if what's happening on Neptune includes another copy of the laboratory with another duplicate Dr. Evil in it. If there's a duplicate on Neptune, Dr. Evil can no longer be sure that Neptune is in fact a distant planet and not the one under his feet (and thus cannot treat it as irrelevant to his probability assignments). As long as the copies of Dr. Evil are unaffected and no new copies are created elsewhere, $P(\text{I'm Dr. Evil}|U)$ should be unaffected by changes to the environment. [Sebens and Carroll, 2018, 40]

Having argued for the plausibility of ESP, Sebens and Carroll argue that its implementation in the context of the EI, which supposedly specifies the self-locating credences of an epistemic agent in the Everettian multiverse, is a quantum version that they call "ESP-QM." That principle states that the probability one should assign to being in a certain branch of an Everettian multiverse does not depend on the full quantum state $|\Psi\rangle$ of the universe but only on the reduced density matrix $\hat{\rho}_{AD}$ that describes the combined system of the agent A and the detector D:

ESP-QM: Suppose that an experiment has just measured observable $\hat{O}$ of system S and registered some eigenvalue O_i on each branch of the wave function. The probability that agent A ought to assign to the detector D having registered O_i [in her branch] when the universal wave function is Ψ, $P(O_i|\Psi)$, only depends on the reduced density matrix of A and D, $\hat{\rho}_{AD}$:

$$P(O_i|\Psi) = P(O_i|\hat{\rho}_{AD}) \tag{10.7}$$

[Sebens and Carroll, 2018, pp. 42f.]

Using an argument idea adapted from Zurek [2005], Sebens and Carroll show that ESP-QM entails the Born Rule as the uniquely correct prescription for assigning self-locating probabilities in the Everettian multiverse. Since, according to them, ESP-QM is the correct Everettian implementation of ESP, which is independently plausible, they regard this result as solving the probability problem of the EI.

Since ESP has a similar motivation as self-locating indifference, which was granted as an assumption to the proponents of multiverse theories in Chapter 7, it seems fine to grant Sebens and Carroll the appeal to ESP as well. But can ESP-QM – the assumption from which the Born rule is in fact derived – really be regarded as nothing more than the appropriate implementation of ESP in the Everettian context? There are strong reasons for doubt.

Let us first investigate how Sebens and Carroll argue that ESP undercuts the supposedly naive application of self-locating indifference in Everettian quantum theory that mandates assigning self-locating credences in alignment with the relative frequencies of branches in which the respective outcomes occur. Sebens and Carroll investigate this question by considering a toy scenario, which they call "Once-or-Twice," in which two agents, Alice and Bob, each have a spin-1/2 particle. Both particles are prepared in the x-spin up eigenstate. First, Alice measures spin in z-direction of her particle, but she does not register the outcome yet. Bob registers Alice's outcome, however, and measures spin in x-direction of his own particle, but only if the outcome of Alice's measurement was $+1/2$. On the Everettian view, branching occurs if and when Bob makes his measurement. Inasmuch as branches have sharp identities, this means that after Bob's measurement, there are now two branches in which Alice's outcome is $+1/2$, but only one branch in which it is $-1/2$. Thus, in this stylized example, at least, branch counting would recommend that Alice ascribe probability 2/3 to being in a $+1/2$ branch, whereas the Born rule recommends ascribing probability 1/2. The verdict derived from branch counting thus differs from that derived from the Born rule, in this case and plausibly many others where branches can be regarded as well defined.

Sebens and Carroll now argue that self-locating reasoning can deliver the Born rule if reasoning in the spirit of ESP is taken into account. They note that the state (reduced density operator ρ_{AD}) of the Alice + Detector system is unchanged when Bob performs his measurement on the branch in which the spin in x-direction of Alice's particle is $+1/2$. They suggest that it seems highly counterintuitive to assume that Alice's probability assessments should change due to an event that does not change the state of the Alice + Detector system. According to them, ESP strongly suggests this conclusion, which allows us to preserve the Born rule, by implying that no part of the universe U that lies outside the set of "internally qualitatively identical states" can influence self locating credences.

To assess whether this is a plausible verdict, let us look at "Once-or-Twice" in a little more detail and consider the situation both before and after Bob makes his measurement. At t_1, immediately after Alice's measurement but and before Bob's, the universal wave function $|\Psi(t_2)\rangle$ is given by

$$|\Psi(t_2)\rangle = 1/\sqrt{2}|\psi_{AD,+}\rangle\,|\chi\rangle + 1/\sqrt{2}|\psi_{AD,-}\rangle\,|\chi\rangle, \tag{10.8}$$

where $|\psi_{AD,+}\rangle$ and $|\psi_{AD,-}\rangle$ are the states of the Alice + Detector system before Alice's registering the measured result in the different branches, and the state $|\chi\rangle$ describes the rest of the universe, including Bob before making his measurement (see eq. (7) in Sebens and Carroll [2018] for more complete decompositions of $|\Psi\rangle$ at all times).

Next, at t_3, after Bob's measurement but prior to Alice's looking at her measurement outcome, the universal wave function is now

$$|\Psi(t_3)\rangle = 1/2|\psi_{AD,+}\rangle\,|\chi_{+,+}\rangle + 1/2|\psi_{AD,+}\rangle\,|\chi_{+,-}\rangle + 1/\sqrt{2}|\psi_{AD,-}\rangle\,|\chi_{-,x}\rangle, \tag{10.9}$$

where the states $|\chi_{+,+}\rangle$, $|\chi_{+,-}\rangle$, and $|\chi_{-,x}\rangle$ are the associated states of the rest of the universe, including Bob.

At t_2, the self-locating problem for Alice seems to be most naturally construed as concerning the system S that consists of her copies in the two branches corresponding to the two states $|\psi_{AD,+}\rangle$ and $|\psi_{AD,-}\rangle$. Now, for ESP to have any relevance with respect to the transition from t_2 to t_3, one would have to make sure that, at t_3, the system with respect to whose subsystems Alice has self-locating uncertainty is still the same subsystem S as before. If not, the self-location problem now concerns a different set of subsystems S', which would mean that ESP is simply silent on how to extract the self locating credences.

The main reason that Sebens and Carroll seem to have for regarding ESP as entailing *no* change in Alice's self-locating credences between t_2 and t_3 is that, as remarked, the reduced density operator of the Alice + Detector system remains the same $|\rho_{AD}\rangle$. But, from the Everettian perspective as endorsed by Sebens and Carroll, this lack of any change in $|\rho_A D\rangle$ leaves out a crucial part of the story – namely, the branching that occurs due to Bob's measurement. Sebens and Carroll are careful to clarify that, as they construe the branching process in Everettian quantum theory, "branching happens throughout the whole wave function" whenever branching happens at all [Sebens and Carroll, 2018, p. 34].[2] They concede explicitly that "[t]he change from t_2 to t_3 in Once-or-Twice increases the number of copies of Alice in existence" [Sebens and Carroll, 2018, p. 46]. But this perspective strongly suggests that Bob's measurement causes Alice to have self-locating uncertainty with respect to a *new* system S' at t_3, which contains more copies of Alice than the system S with respect to which she has self-locating uncertainty at t_2. In other words, it seems plausible that, by Sebens and Carroll's own standards, Bob's experiment *does* affect the system concerning which Alice has self-locating uncertainty and is therefore *not* declared irrelevant to Alice's self-locating credences by ESP.

These considerations assume that branch number is a well-defined quantity, and the Everettian, as observed before, may object to them by arguing that branch identities are, in general, fuzzy. But this objection unfairly shifts the burden of proof in favor of the Everettian. Inasmuch as branch number is not defined due to branch fuzziness, this creates additional problems for her analysis because it makes the very meaning of self-location in Everettian quantum theory unclear.

Moreover, at least in an approximation where branches are (to some sufficient degree of precision) well-defined and precise branch numbers with specific Born weights exist, any suggested solution to the probability problem, in order to be successful, must produce results that reproduce the empirically confirmed verdicts. Branch fuzziness makes the probability problem harder to solve for the Everettian, not easier. Inasmuch as it may appear to make it easier, it does so only by muddying the waters.

To conclude, if Alice's self-locating uncertainty at t_3 concerns a different system S' rather than the system S that is relevant at t_2 in the setting where branch numbers are precisely defined, we should assume that this is the case also in the fuzzier setting where branch numbers are no longer precisely defined.

Since the system with respect to which Alice has self-locating uncertainty changes from t_2 to t_3, ESP is simply silent with respect to whether Alice's credences should change. Notably, it does not entail anything about whether, at t_3, Alice should rely on the Born rule or, say, branch counting when assigning her self-locating credences.

In contrast, to ESP, however, ESP-QM does have a bearing on the development of Alice's rational credences when Bob makes his measurement – namely, as shown by Sebens and Carroll, that they should conform to the Born rule at all times. This, however, shows that ESP-QM has consequences that go beyond those of ESP and

[2] This view is criticised as implausible by McQueen and Vaidman [2019], according to whom Sebens and Carroll fail to consider cases of genuine self-locating ignorance. McQueen and Vaidman suggest an alternative derivation of the Born rule from principles of self-locating belief, which would deserve a separate critical analysis.

so is, contrary to Sebens and Carroll, not merely "a less general version of ESP" [Sebens and Carroll, 2018, p. 42] tailored to the Everettian context. Notably, for the derivation of the Born rule from ESP-QM to have the significance that it has according to Sebens and Carroll, an independent motivation for ESP-QM would be needed, one that goes beyond the motivation provided for ESP.

In the absence of any such a motivation – and in the absence of an alternative, independent physical derivation of the Born rule from the core tenets of Everettian quantum theory – assuming ESP-QM to derive the Born rule amounts to nothing else than, as in Vaidman's early account, assuming it from the start.

There are several further extant attempts to motivate the Born rule from the branching structure of the Everettian multiverse beside the ones considered here. At least those, however, seem to have serious shortcomings. If we add to this picture the difficulties in defining the branching structure reviewed in the previous section, the overall outlook for the Everett interpretation seems bleak. The two multiverse views to be discussed in the next two sections – David Lewis's modal realism and Max Tegmark's level IV multiverse of mathematical structures – are plagued by similar problems.

10.2 Modal Realism

David Lewis's *modal realism* [Lewis, 1986b] is the astounding hypothesis that all worlds that are "metaphysically possible" really exist. It is worthwhile to devote a section of this chapter to modal realism because it can be seen as a multiverse theory, albeit one motivated by philosophical, not physical, considerations. Modal realism is central to Lewis's overall philosophical system, but despite Lewis's fame, it has never become an extremely popular metaphysical view and does not seem to have many adherents today. Nevertheless, it is rewarding to consider the repercussions of the considerations on multiverse theory testing developed in this book for the assessment of modal realism. The outcome of that assessment will be devastating: to the extent that modal realism is intelligible at all, it is not a position that one could reasonably believe if one takes it really seriously.

10.2.1 Why Modal Realism at All?

The starting point of modal realism is our everyday use of "modal" language, i.e., language involving modal notions such as "possible," "necessary," "contingent," and "impossible." Due to the influential work of philosopher Saul Kripke [1972], it has become a widely held view among philosophers that it is helpful to employ a separate notion of *metaphysical modality* – i.e., metaphysical possibility and necessity – in addition to notions of linguistic modality – analytic and synthetic truth – and epistemic modality – a priori and a posteriori knowledge – whose importance had been more widely appreciated when Kripke put forward his arguments.

A stock example of a sentence with an interesting modal profile is "Water is H_2O." According to Kripke, this sentence is metaphysically necessary – the stuff that we denote by "water" is H_2O – but it is both linguistically and epistemically contingent: both "water" and "H_2O" rigidly denote the same substance, so it is not metaphysically possible for "Water is H_2O" to be false; at the same time, the identity of water and H_2O had to be discovered through the use of science, so it is neither part of the meaning of "water" that this expression denotes H_2O nor is the sentence an a priori truth whose truth we could, in principle, have discovered by purely armchair philosophical consideration. Not all philosophers agree with Kripke on this analysis, but the distinctions highlighted by these considerations are widely accepted as useful and relevant.

Many philosophers have found it convenient to frame talk about (metaphysical) possibility using the jargon of "possible worlds." In that jargon, that a proposition holds with metaphysical necessity is expressed by saying that it holds in all (metaphysically) possible worlds. Analogously, that it holds with (metaphysically) contingence is expressed by saying that it holds in at least some (metaphysically) possible worlds. Quantification over possible worlds thus allows one to formalize propositions that ascribe metaphysical necessity and contingence to states of affairs using only the simple and elegant language of first-order logic rather than some more contentious formalism of modal logic.

Most philosophers interpret talk about "possible worlds" as nothing more than jargon. In other words, they regard reference to possible worlds as a sometimes convenient manner of speaking but not as entailing commitment to the literal *existence* of other possible worlds beyond our own. David Lewis's modal realism arises from disagreement with this cavalier attitude toward possible world jargon. According to Lewis, the language of possible worlds is such a powerful theoretical tool that, being intellectually honest, we should take our commitment to it seriously and embrace *modal realism*: the astounding view that all metaphysically possible worlds are real.

Lewis sees four main areas in which reference to possible worlds has significant advantages [Lewis, 1986b, sections 1.2–1.5]. First, as already said, reference to possible worlds streamlines modal discourse. Second, reference to possible worlds enables an elegant understanding of *counterfactual conditionals* (sentences of the form "If it had [not] been the case that . . . , it would [not] have been the case that. . .), which in turn are useful to spell out the meaning of *causal* statements. For example, the causal statement "The extinction of the dinosaurs was caused by some giant asteroid which hit the earth" can be (somewhat roughly) paraphrased as "If some giant asteroid had not hit the earth, the dinosaurs would not have become extinct," which has the form of a counterfactual conditional. Lewis proposes to construe counterfactual conditionals in terms of a similarity relation among possible worlds. According to that analysis, the counterfactual conditional about the extinction of the dinosaurs just stated is true if and only if in the possible worlds that are most similar to our own but that do not feature an asteroid hitting the earth at the appropriate time, the dinosaurs do not become extinct. Third, according to Lewis, both semantic and mental *content* are most easily systematized by appeal to possible worlds, and fourth, quantification over *properties* can be elegantly formalized using it.

Is it possible to interpret quantification over possible worlds figuratively – i.e., as not committing the speaker to the literal existence of all the possible worlds? Lewis considers various options – "linguistic ersatzism," "pictorial ersatzism," and "magical ersatzism" – and finds them all seriously problematic in different ways [Lewis, 1986b, chapter 3]. He concludes that the most straightforward interpretation of quantification over possible worlds, which involves ontological commitment to all those possible worlds; i.e., the endorsement of modal realism, is more appealing than any of the different types of ersatzism.

Whether the jargon of possible worlds indeed merits being taken as seriously as Lewis argues seems questionable, however. That jargon may be helpful for certain purposes to systematize and streamline modal discourse, but to hypothesize possible worlds as if they were scientific objects comparable to, say, electrons or genes runs the risk of distracting our attention from the social and psychological mechanisms that actually govern our use of modal concepts. For example, as just mentioned, Lewis appeals to an unwieldy notion of similarity among possible worlds to obtain an analysis of causal counterfactuals that recovers our intuitive verdicts about cause/effect relations. But that notion has no independent motivation and does not do justice to the importance of *intervention* and *manipulation* that more recent accounts of causation highlight (e.g., Woodward [2003]).

10.2.2 Could We Seriously Believe It?

Is modal realism coherent in the first place, and, to the extent that it is, can we coherently believe it and act in accordance with this belief in our everyday lives? Forrest and Armstrong argue that modal realism is flat-out inconsistent because it leads to contradictions analogous to those that arise in naive set theory [Forrest and Armstrong, 1984]. They assume that, for all things that exist in any possible worlds, there is a possible world that contains exact copies of all those things. Accordingly, if we assume a totality of all possible worlds, there will also be a possible world that contains them all. But this world would have to be, in a strict sense, larger than all possible worlds and, hence, cannot possibly exist, in contradiction to the initial assumption that it is a possible world itself. In response to this challenge, Lewis resorts to a proviso to the principle that things from different possible worlds can be combined into a single possible world, which undercuts the inferences drawn by Forrest and Armstrong [Lewis, 1986b, section 2.2].

The idea of a totality of – existing – possible worlds raises further questions. One difficulty is that it seems difficult to determine whether there are any common features of all metaphysically possible and, if so, which. For example, according to Lewis, any two objects in a single possible world are spatiotemporally related (a thesis that he himself does not feel completely comfortable with [Lewis, 1986b, p. 71]). But this view presupposes that spatiotemporal categories apply to the objects of all possible worlds, which in turn rules out – questionably, one may feel – the mere *metaphysical possibility* of worlds without space and time. But the main problem for Lewis's position is not that this presupposition seems questionable – i.e., that worlds without space and time seem intuitively possible. The main problem for Lewis's position is the fact that we have nothing more than very shaky intuitions to appeal to when making judgments of that kind.

If we concede to Lewis, at least for the sake of the argument, that modal realism is at least internally consistent, we can ask how it fares, as a multiverse theory, when we subject it to the kind of empirical scrutiny that, along the lines sketched in Chapter 7, we would like to apply to multiverse theories in physics. An immediate worry about modal realism's multiverse of possible worlds is that agents in many other possible worlds are deceived by their experience and their surroundings such that standard strategies of rational inference will lead them to wrong conclusions. For instance, many other possible worlds harbor observers whose memories are "fake" (similar to those of hypothetical Boltzmann brains as considered in Section 8.3). Many harbor observers whose expectations concerning their own future – whether based on "fake" or genuine memories – are prone to be radically disappointed in that their future will not follow the regularities of their (actual and/or remembered) past. (After all, as we learned from Hume, there is no logical guarantee that past regularities carry over to the future, so, in some possible worlds, the future will be radically different from the past.)

Indeed, the worry that seriously embracing modal realism would commit one to endorsing inductive skepticism – i.e., radical skepticism concerning whether past regularities carry over to the future – is one of the earliest objections against modal realism, pioneered by Forrest [1982] and further developed by Thomas [1993]. (Lewis [1986b, p. 116] credits also George Schlesinger, Robert M. Adams, and J. J. C. Smart with it.) The idea that underlies this worry is simple: if there are many observers across worlds whose psychological states are exactly like ours, but in whose future inductive reasoning breaks down, we have no reason any more to be confident that induction will reliably work in our own future because we have to seriously take into account the possibility that we might be such a "pathological" observer in a "pathological" world.

Lewis himself professes to be unconcerned by this problem, mainly because he does not see it as posing any additional difficulties over and above the familiar problem of induction as pointed out by Hume:

I, as a modal realist, have no more reason to foresake inductive reason than anyone else has. ...By trusting induction we run a risk, and we proceed in the confident hope that the genuine possibilities or error will seldom be realised. All that, I say, is quite independent of any theory of the nature of possibilities. I recognise the possibilities of error that everyone else recognises; they are no more and no less possibilities of error for being understood as other worlds of a kind with our own. [Lewis, 2001, p. 121]

Lewis then offers some speculations about why it may intuitively seem – mistakenly, as he argues – that inductive reasoning is more problematic if other possible worlds exist than if they do not exist.

However, Lewis's claim that acknowledging the reality of other possible worlds does not change the status of induction is unconvincing. To see this, imagine that, like Dr. Evil in the scenario considered by Sebens and Carroll, you learn about the existence of a number of people with exactly the same psychological states and memories as yourself, where for only some of those people their memories more or less faithfully reflect their personal histories whereas for the others their memories are as misleading and "fake" as they are for a Boltzmann brain. Prima facie it would then not be reasonable for you to trust your memories – unless you *additionally* learned that the number of observers with fake memories who are copies of you is vanishingly small compared with that of observers with reliable memories who are copies of you.

The lesson from these considerations for the assessment of modal realism is straightforward: many possible worlds are inhabited predominantly by observers with fake memories and/or futures where inductive reasoning breaks down. Unless Lewis provides reasons to believe that, if modal realism is true, those pathological observers are, in some relevant sense, a tiny minority, believing in modal realism is impossible: all our empirical knowledge, including the knowledge encoded in modern physics, is based on the assumption that our memories are not "fake" and, in our everyday lives, we rely on the expectation that the future will in some respects resemble the past. If modal realism makes that expectation unreasonable, it is not coherently believable.

According to Forrest [1982], modal realism leads to inductive skepticism because, across possible worlds, observers in pathological worlds where memories are "fake" and induction breaks down outnumber regular observers by any reasonable standard. Lewis disputes this in [Lewis, 1986b, section 2.5] by arguing that cardinalities of pathological versus regular observers are not well defined. But this reply is inadequate because it unfairly shifts the burden of proof: it is Lewis who postulates the existence of pathological observers, so it is Lewis who owes us reasons to be confident that we are not pathological even though, evidentially, we might be.

Lewis's style of philosophizing is in some respects inspired by the empirical sciences: according to him, since reference to possible worlds comes with certain theoretical benefits, committing oneself to their existence is legitimate (or, actually, required) in a similar way to how committing oneself to the existence of electrons is legitimate because of the theoretical (and empirical) benefits that come from accepting quantum electrodynamics. There is nothing wrong with learning from science when doing philosophy. But if Lewis infers the truth of modal realism like a physicist would infer the truth of some physical theories, he must also accept its confrontation with experience as if it were a usual multiverse theory. As discussed in Chapter 7, it is a widely acknowledged minimal requirement for such a theory to qualify as empirically adequate that our observations are (at least somewhat) typical if it is true. Insisting, as Lewis does, that we can trust in being non-pathological observers but also insisting on an abundance of real pathological observers – without giving us any reason to think that non-pathological observers are in some relevant sense highly atypical among inhabitants of all possible worlds – means not taking his own modal realism as seriously as he should take it when advancing it as a serious philosophical theory of modality.

10.3 A Multiverse of Mathematical Structures?

10.3.1 Tegmark's Level IV Multiverse

Max Tegmark's hypothesis of the level IV multiverse of mathematical structures rivals David Lewis's modal realism in terms of extravagance. It states that all mathematical structures are physically realized as distinct universes and together form a multiverse that goes far beyond level II (cosmological) multiverses such as the landscape multiverse and the level III multiverse of Everettian quantum theory discussed in Section 10.1.

The universes that supposedly populate the level IV multiverse are incredibly diverse. As Tegmark explains, some of those universes are similar to our own in that they realize laws similar to those encoded in the Standard Model of elementary particle physics, but with other gauge groups. Assuming that string theory is mathematically consistent, whether or not it correctly describes *our universe*, it is guaranteed that the level IV multiverse contains universes in which string theory is, as hoped by its proponents in this universe, the "correct Theory of Everything."

In the level IV multiverse, things do not stop at the level of other gauge groups and string theory. As Tegmark puts it, "[w]hen contemplating the Level IV multiverse, we need to let our imagination fly, unencumbered by our preconceptions of what laws of physics are supposed to be like" [Tegmark, 2014, p. 324f.]. For example, as he explains, there are universes in the level IV multiverse in which space-time is discrete and even universes without any space and time (because these are the universes that realize mathematical structures that are completely different from anything resembling space and time).

10.3.2 Universes Are Not Mathematical Structures

Tegmark's central reason for believing that there is indeed a level IV multiverse is that, as he sees it, mathematical and physical existence are ultimately equivalent. Notably, as he argues, the physical world (whether a single universe or a multiverse) *is itself* mathematical in that, as he puts it, "[o]ur external physical reality is a mathematical structure"[Tegmark, 2014, p. 254] – a claim that he dubs the *Mathematical Universe Hypothesis* (MUH). Tegmark offers two arguments for MUH, which, however, fall dramatically short of establishing the MUH. Since Tegmark's case for the view that mathematical and physical existence are equivalent is entirely built on the MUH, undermining his arguments for MUH amounts to eliminating the reasons that he gives for belief in the level IV multiverse.

Tegmark's first argument for the view that our own universe is a mathematical structure goes as follows: if our universe has objective reality "completely independent of us humans" [Tegmark, 2014, p. 254], then a complete description of it has to be free from human "baggage." Furthermore, according to him, all empirical concepts that we use – including *"protons, atoms, molecules, cells,* and *stars"* – are created by us, which means that a description in terms of them may not be intelligible to nonhuman agents such as *"aliens or supercomputers"* [Tegmark, 2014, p. 255]. Consequently, as he sees it, any description in terms of such concepts would *not* be baggage-free and, consequently, could not possibly be a complete description of objective reality. So, unless we are prepared to give up the idea of objective reality altogether, we must accept that a complete description of it will only use purely mathematical terms. According to Tegmark, this means that what any hypothetical complete description of reality would describe – i.e., objective reality itself – must inevitably be as much a mathematical structure as a physical structure. Thus, according to Tegmark, if there is an objective reality at all, it must be a mathematical structure; the MUH holds.

Tegmark seems right that there is some sense in which a complete description of objective reality – inasmuch as the idea of such a description is meaningful – would have to be "baggage-free." For instance, a description in terms of electromagnetic potentials A_μ (assuming that it were accurate) would not be "baggage-free" in that two such descriptions that differ only by the addition of some scalar field gradient $\partial_\mu \Lambda$ are physically identical. Configurations of A_μ do not correspond one to one to physical states of affairs, which means that descriptions in terms of potentials A_μ include some arbitrariness that one may regard as "baggage."

But is it plausible that a description of physical reality, in order to be objective and complete, would have to be free from *any* empirical vocabulary, which – Tegmark argues – is *created* by us, unlike mathematical vocabulary? Almost the exact opposite seems to be the case: in order for something to qualify as a – potentially accurate – description of physical reality in the first place, a linguistic understanding of it must be in place that allows one to connect certain aspects of it to empirical phenomena. Only then can we potentially use it for the purposes of description, explanation, and manipulation. "Aliens and supercomputers," to use Tegmark's examples, might find other empirical concepts more natural and straightforward to use than we do, and this may indeed make the idea of a completely objective description of physical reality more intricate and problematic than it initially seems. However, purely mathematical considerations, which do not contain any empirical vocabulary, do not by themselves apply to physical reality at all (though one frequently can apply them, often in multiple, non-unique ways) and are not candidate descriptions of physical reality in the first place.

In fact, in order to express the point that the electromagnetic potential A_μ contains "baggage" in the sense of "surplus formal structure," we must unavoidably appeal to a notion of physical existence as distinct from mathematical existence: by saying that different configurations of A_μ that differ only by a scalar field gradient $\partial_\mu \Lambda$ are physically identical, we are not making a mathematical statement but one that pertains to the correct empirical interpretation of the mathematical formalism. By equating physical existence with mathematical existence, we would rob ourselves of the ability to express this and similar statements. That such statements are important to clarify the objective empirical content of a physical theory shows that the MUH cannot possibly be correct.

What about Tegmark's second argument for the MUH? The core idea of that argument is that an objective description of physical reality, in order to be complete, would have to be *isomorphic* to physical reality itself.[3] But if two structures are isomorphic, then they satisfy the same – appropriately interpreted – propositions and

---

[3] Tegmark does not always distinguish between equations that express a universe's laws of nature and equations that provide a complete description of that universe. For example, he writes that the "equations [of the coveted Theory of Everything] are a complete description of the mathematical structure that is the external physical reality" [Tegmark, 2014, p. 255]. But unless one assumes that a universe's laws of nature completely determine the physical facts that obtain in it, the laws that govern a universe and a complete description of that universe should be kept apart. Given the more blatant difficulties with Tegmark's position outlined here, I otherwise ignore this complicating factor.

can, for the purposes of mathematics, be regarded as identical. According to Tegmark, it follows that a physical structure and a mathematical structure isomorphic to it, which in turn is completely described by some mathematical theory, are not merely equivalent in the sense of isomorphism but literally identical, establishing the Mathematical Universe Hypothesis.

Tegmark is correct that, in mathematics, one can regard two structures as, for relevant purposes, identical if there is a one-to-one correspondence between them. For example, one can regard the real numbers as identical either to equivalence classes of Cauchy sequences of rational numbers or to Dedekind cuts. The set of real numbers is isomorphic to both those sets.

But even within mathematics, the claim that isomorphism amounts to identity must be handled with care. The claim that the real numbers are identical to equivalence classes of Cauchy sequences and/or to Dedekind cuts must be understood in the right way. Notably, whether one treats equivalence classes of Cauchy sequences of rational numbers or Dedekind cuts as identical to the real numbers is a matter of convention. Both these structures (alongside various others) can "play the role" of the real number structure. This differs from mathematical identity facts such as that the successor of 1 and the smallest prime number are identical, which are dictated by the axioms of (in this case) arithmetic, combined with the laws of logic.

Outside of mathematics, the claim that isomorphism boils down to identity is of course plain false: two physical systems that are described by the same mathematical structure can be radically different; one may think of, for example, a liquid and a fundamental classical field that are governed by the same differential equation and realize the same solutions of that equation. And if we compare a physical and a mathematical structure, as we have seen in the answer to Tegmark's first argument, there is an irreducible need for empirical vocabulary to identify the objective features of the physical structure, and this vocabulary, of course, has no place in the description of the mathematical structure. To conclude, where some physical structure is equivalent to some mathematical structure in the sense of isomorphism, identity between the two does not follow. Tegmark's arguments for the Mathematical Universe Hypothesis fail, and with them, his reasons to believe that mathematical and physical existence are the same.[4]

---

[4] I have argued elsewhere [Friederich, 2011] that, in mathematics, language does not work in the descriptive, fact-stating way in which it is usually deployed in sciences such as physics. In my view, when mathematics is pursued as pure mathematics – i.e., without an eye to specific extra-mathematical applications – mathematical sentences are better thought of as norms that govern the use of the concepts they contain than as descriptions of any independent matters of fact. The main motivation for this view is that, according to the modern (Hilbertian) understanding of axiomatics, the axioms of a mathematical theory are implicit definitions of the concepts they contain, which means that, if they are consistent, the question of whether they are true or false according to some external standard does not make any sense. This makes them radically unlike sentences about the physical world and indicates that mathematical existence differs profoundly from physical existence. For the reasons provided in the main text, however, subscribing to this view is not required to find fault with Tegmark's arguments for the MUH.

Without this belief, in turn, the very idea of a level IV multiverse loses whatever appeal it may have had. The thought that all mathematical structures, taken together, form a physical super-multiverse is not even coherent.

To conclude, neither the Everettian multiverse nor David Lewis's multiverse of possible worlds nor Max Tegmark's thesis of the mathematical multiverse can be seriously believed. Conceptually and epistemologically they run into insurmountable problems.

11

Outlook

11.1 Is the Multiverse Even Science?

Throughout the discussion of multiverse theories in this book, I have set aside the most heated discussions about the multiverse and ignored the most scathing criticism of such theories: that the impossibility of probing the properties of the suggested other universes through more or less direct empirical access makes theories that entail their existence and ascribe various properties to them unscientific.

Peter Woit, a vocal critic of "multiverse mania" in print [Woit, 2006] and on the Internet, is perhaps the most outspoken public proponent of this view. According to him, the hope of ever finding compelling support for a theory that entails the existence of other universes by means of observations confined to our own universe is naive. As he puts it in his blog in response to the question "Why do you describe the Multiverse as 'pseudo-science'?":

Since you can't observe anything about it directly, the multiverse must be justified in terms of another theory that can be tested and this is string theory. But if you talk to string theorists these days about how they're going to test the unified theory that string theory is supposed to provide, their answer is that, alas, there is no way to do this, because of the multiverse. You see, the multiverse implies that all the things you would think that string theory might be able to predict turn out to be unpredictable local environmental accidents.

So, the multiverse can't be tested, but we should believe in it since it's an implication of string theory, but string theory can't be tested because of the multiverse.[1]

As we have seen in Chapter 7, a formalism can be developed that makes multiverse theories testable at least in principle, even if they entail universes that are vast, heterogeneous, and very different from each other. For multiverse theories that

---

[1] Quoted from Woit's blog, www.math.columbia.edu/$\sim$woit/wordpress/?\discretionary-wp$\_$super$\_$ $f\discretionary-aq=why-do-you-describe-the-multiverse-as-pseudo-science, accessed August 21, 2019.

entail a multiverse that is uniform in many ways – i.e., a multiverse that in many respects behaves like a single universe – testability may actually be quite good.

On the other hand, the discussion in Chapter 8 highlighted serious, possibly devastating problems that arise in concrete attempts to make specific multiverse theories testable. That discussion provides grist to Woit's mill. It may be, in principle, possible to extract concrete nontrivial predictions from specific multiverse theories using, notably, considerations about self-locating belief. But the researcher degrees of freedom that arise in the process – e.g., when making choices about observer proxy, cosmic measure, cosmic background conditions, are prone to be affected by confirmation bias – which makes such predictions hopelessly unreliable.

Does this vindicate Woit's wholesale dismissal of multiverse theories as pseudoscience that physicists should best ignore? I don't think so. The problems that arise in attempts to extract concrete predictions from multiverse theories in large part have to do with contingent, human, epistemic liabilities like the tendency to fall prey to confirmation bias. There is no reason to believe that we are *in principle* unable to establish the truth of a multiverse theory if that theory is true. In principle, it is entirely possible that we come to rationally – and, if there is a multiverse, correctly – believe that we live in a multiverse – namely, if we have a physical theory T that makes some striking general empirical predictions, which end up confirmed, and, at the same time, entails that there are other universes where different values of certain parameters are realized. The new fine-tuning argument for the multiverse as developed in Chapter 6 – or, for those who accept it, the standard argument as discussed in Chapters 4 and 5 – may be seen as providing an additional reason for taking multiverse theories especially seriously in the first place.

But what the present discussion suggests is that, in *practice* rather than principle, the pitfalls in obtaining compelling evidence for or against such theories are enormous.

In fact, Woit is probably correct about the *sociological* risks associated with great enthusiasm about multiverse theories. In view of the researcher degrees of freedom waiting to be exploited in attempts to extract concrete predictions from specific multiverse theories – e.g., those aimed at deriving likely values of the cosmological constant in the landscape multiverse scenario – investing great efforts in such attempts will likely be a waste of time and resources. Even more detrimentally, exploiting those researcher degrees of freedom can be used as an effective – though not consciously deployed – strategy to create the misleading impression that some research program is empirically fruitful. According to Woit, this is precisely the problem with string theory. Whether or not he is correct here, the problems with testing multiverse theories may make it methodologically wise to invest somewhat more time and resources in investigations of theoretical alternatives, even if there are reasons to believe that some multiverse theory might be correct.

11.2 Whither Physics?

A sobering conclusion is entailed by the combination of two of the views defended in this book: namely, on the one hand, that we should seriously consider the possibility that certain seemingly fundamental parameters of physics might be environmental – i.e., that we might live in some type of multiverse – and, on the other hand, that performing conclusive tests of specific multiverse theories that yield compelling verdicts about their truth or falsity will likely remain extremely difficult, perhaps impossible, in practice. Taken together, these two views suggest that we may have to arrange ourselves, perhaps permanently, in a situation where we simply do not know whether there are other universes and, if there are, what they are like.

If it is true that we may never – or at least not for a very long time – have strong reasons for or against believing that there are other universes with different parameters, this would in itself indicate a serious boundary to our physical knowledge. But we should also expect it to indirectly make progress in fundamental physics much more difficult. Typically, such progress occurs because problems and anomalies – either empirical or theoretical – are identified for our currently best theories of fundamental physics. If those problems are not solvable in the context of those theories, a successful solution to them inevitably, almost by definition, means a major theoretical breakthrough. But permanent uncertainty about the existence of other universes with different values of the parameters also creates uncertainty about what constitutes a major problem of fundamental physics in the first place – i.e., a problem that can potentially be solved with the benefit of a major theoretical breakthrough.

Notably, as we saw in Chapter 4, attempts to account for the distribution of the values of fundamental parameters in our universe are futile if we live in a multiverse where those parameters "scan" across universes. As I argued there, if we were confident of the truth of such a multiverse theory on other grounds, we could either regard the values of those parameters in our universe as being unsurprising because the parameters had to be right for life somewhere and we could not have found ourselves anywhere else, or we could regard those values as primitive coincidences that admit no further explanation. This difficulty makes it unclear whether, for instance, the pattern of quark and lepton masses in our universes should be seen as a problem with a potentially interesting solution; similarly, it makes it doubtful whether the naturalness problem of the Higgs mass and the cosmological constant problem should be seen as keys to important advancements in fundamental physics.

Because string theory is one of the leading candidate theories for unifying elementary particle physics and gravity and string theory, according to many, must be considered in the shape of the landscape multiverse, our persisting inability to conclusively test multiverse theories may even detrimentally affect our ability to make progress on the general fundamental problem of combining particle physics and gravity in a unified fundamental theory.

In a recent, rightly celebrated book, Sabine Hossenfelder [2014] documents her critical questioning of the crisis into which modern fundamental physics appears to have maneuvered itself. She explores the suspicion that the ideal of mathematical beauty, as encoded in the naturalness criterion, has impeded physical progress in the past few decades. Her diagnosis may well be correct that it is very risky to search for a theory that above all other things is as beautiful as possible by the theorists' historically grown standards.

The considerations developed in this book suggest another, perhaps complementary diagnosis of what lies at the heart of the current crisis and lack of progress in fundamental physics. That diagnosis is that we may have entered an era where we must take the possibility seriously that we might live in a multiverse and that whether or not this possibility holds affects what we regards as a solvable problem in the first place. Notably, we are confronted with the problem that we simply do not know whether some of the parameters in our currently best (and most promising candidate) fundamental physical theories are environmental and, if so, which.

Whether, for example, the cosmological constant is environmental or not matters greatly for whether looking for a mechanism or principle that accounts for its measured value is a promising research strategy. The same holds for the hierarchy between the Planck mass and the scale of electroweak symmetry breaking. Attempts to account for the values of the cosmological constant and the electroweak scale are promising only if these quantities are *not* environmental. If they are environmental, there might just not be anything to explain in these cases.

On the bright side, there are pressing problems in fundamental physics that may well hold the key to new theoretical breakthroughs and to which the question of whether we live in a multiverse seems largely irrelevant. Two very different such problems that seem both very promising are as follows:

First, there is the dark matter problem. There is strong evidence via various different, partly independent channels of evidence that the majority of matter in the visible universe is not made up of any known type of particle. The first hint for the existence of that matter was that, if it were not there, rotating galaxies would rapidly disintegrate, given their observed velocity of rotation [Corbelli and Salucci, 2000]. In the meantime, further independent lines of (partly indirect) evidence for such additional matter have emerged – for example, from gravitational lensing, from the cosmic microwave background, and from cosmic structure formation [Bertone et al., 2005].

Because that additional matter, whose existence is inferred, interacts with known types of matter mostly through gravitation (and possibly through the weak interaction) but not via electromagnetism, which describes the properties of light, it is commonly referred to as "dark matter." Many theories have been proposed as to what type of particle dark matter might consist of, but none of them has been compellingly supported by observations. All attempts at direct detection of dark matter particles so far have been unsuccessful [Bertone and Tait, 2018].

The existence of dark matter is rather well supported, but no known type of particle whose existence is empirically confirmed is a candidate dark matter particle. Identifying the actual dark matter constituents is a major open problem in fundamental physics. Since there is no candidate dark matter particle in the Standard Model of elementary particle physics, it cannot be solved within the Standard Model framework. Solving it would thus be virtually guaranteed to come hand in hand with a major advancement in our understanding of fundamental physics.

Since the evidence for dark matter largely relates to its gravitational effects, there is also the possibility that the correct answer to the dark matter problem also requires revising our views of gravitation – a possibility advocated, for instance, by Kroupa et al. [2012]. The best-known such models are MOND (modified Newtonian gravity) [Milgrom, 1983] and its relativistic generalization TeVeS [Bekenstein, 2004].

Unlike the dark energy problem, which is associated with the cosmological constant, the challenge of identifying the nature of dark matter is largely independent of the question of whether we live in a multiverse (though assuming a multiverse setting may make it more difficult to evaluate the empirical consequences of specific dark matter theories [Azhar, 2014] and researcher degrees of freedom abound in such attempts). If we indeed remain ignorant for a long time – perhaps permanently – about whether there are other universes with different parameters, this will not make identifying the true nature of dark matter any less valuable, nor will it reduce the expected progress we are likely to make if we manage to solve the dark matter problem.

The second problem is accounting for how reality can possibly be such that quantum theory has the success that it manifestly has. In Section 10.1, it was outlined why the continuing unrivaled empirical success of the quantum theoretical framework for formulating physical theories is mysterious. For several decades, the field had been dominated by the Copenhagen interpretation, which unfortunately is elusive and unclear. One of its few clear-cut tenets, however, is that, as a framework for fundamental physical theories, quantum theory is fine as it stands and not in need of being complemented by an account that clarifies what the "ontic variables" really are. Thus, according to the Copenhagen interpretation, the foundations of quantum theory are not an area of research in which major breakthroughs are to be expected.

More recent, broadly "neo-Copenhagenian" approaches like Richard Healey's pragmatist interpretation [Healey, 2017], quantum Bayesianism (striving for an information-theoretic reformulation of quantum theory) [Fuchs and Schack, 2015], and my own "therapeutic" approach [Friederich, 2014] share this tenet. It is unclear, however, whether these approaches are really able to demystify the success of quantum theory. Neither of them offers a clear-cut criterion of which variables denote objective physical quantities and which do not. This makes it difficult to see how they can possibly account for the objectivity of everyday macroscopic facts in a coherent way without drawing an arbitrary and ad hoc line between an unspeakable micro- and an objective macro-level or without otherwise relying on primitive anthropocentric notions in a problematic way.

The only "realist" approach to quantum theory that does not propose any amendments to or adjustments of the formalism of the theory is the Everett interpretation. We have seen in Section 10.1 that it runs into extremely serious difficulties, concerning both its metaphysics and its epistemology.

One important worry about approaches to the foundations of quantum theory that make amendments or adjustments to the theory is that they might not convey any additional predictive power to the theory. This worry is serious, especially in view of a theorem by Colbeck and Renner [2011], refined by Leegwater [2016], according to which no theory that reproduces the empirical consequences of quantum theory and has a prima facie attractive feature called *parameter independence* can have superior predictive power compared to quantum theory itself.

Still, if approaches that do not make any amendments or adjustments to quantum theory are unable to account for the theory's empirical success or are not even coherent, we have simply no choice: we are left with approaches that do make such amendments or adjustments and thus require a significant overhaul of the quantum theoretical framework. Extant such approaches include pilot wave theory and the GRW model, but there are many more ideas for such approaches.

Perhaps most intriguingly, there are arguments in favour of accounts of quantum theory that are most straightforwardly conceptualized in terms of causal influences backward in time. Whether or not this idea of·"retrocausality" in the foundations of quantum theory is as fruitful as its advocates such as Price [1996] hope remains to be seen. (For an overview of the motivation, problems, and concrete versions of retrocausal accounts in the foundations of quantum theory, see the review article [Friederich and Evans, 2019], which I co-wrote.)

In any case, one can view the problems with neo-Copenhagenian and Everettian accounts mentioned as an indication that progress in the foundations of quantum theory might only be possible by revising core assumptions concerning what we expect from fundamental physical theories. And if any revision of such a core assumption – say, the absence of causal influences backward in time – were universally accepted, whether based on empirical or predominantly systematic grounds, this would, no doubt, be regarded as a major breakthrough in the overall foundations of physics.

The overarching lesson of the considerations developed in this book arguably is a humbling one. Hilbert's dictum, "We must know – we will know," inscribed on his tomb, might just be wrong when it comes to fundamental physics. One may already find it implausible when one reflects on the possibility, argued for by McKenzie [2011], that there might simply not be any *ultimately fundamental* physical micro-level whose laws we might ever come to understand. There is simply no guarantee that the hierarchy of laws of physics ever bottoms out, and even if it does so at some point, there is no guarantee that it does so at a level sufficiently close to everyday scales such that we humans – or any superintelligent technology-created intellectual "descendants" of us – can ever obtain epistemic access to it.

Our knowledge about fundamental physics might forever remain incomplete not only "in the small." The considerations developed in this book suggest that it may also forever remain incomplete "in the large." Notably, we may never really know whether there are other universes with different parameters and, if so, what the properties of those universes really are. Progress in fundamental physics will continue, hopefully including in the fields of dark matter research and the foundations of quantum theory. But if a large portion of the features of reality to which we have more or less direct causal access, including central aspects of the laws in the part of reality to which we have empirical access, are environmental rather than universal, our knowledge of fundamental physics may have severe permanent limits. Our well-confirmed views of reality may forever be constrained to a tiny bit of something far more vast and far more diverse about which we can merely speculate.

References

Adams, F. C. Stars in other universes: Stellar structure with different fundamental constants. *Journal of Cosmology and Astroparticle Physics*, 08:10, 2008.

Adams, F. C. The degree of fine-tuning in our universe – and others. *Physics Reports*, 807:1–111, 2019.

Adams, F. C., and Grohs, E. On the habitability of universes without stable deuterium. *Astroparticle Physics*, 91:90–104, 2017.

Adlam, E. The problem of confirmation in the Everett interpretation. *Studies in History and Philosophy of Modern Physics*, 47:21–32, 2014.

Aguirre, A., and Johnson, M. C. A status report on the observability of cosmic bubble collisions. *Reports on Progress in Physics*, 74:074901, 2011.

Aguirre, A., and Tegmark, M. Multiple universes, cosmic coincidences and other dark matters. *Journal of Cosmology and Astroparticle Physics*, 2005:003, 2005.

Albert, D., and Loewer, B. Interpreting the many worlds interpretation. *Synthese*, 77:195–213, 1988.

Albrecht, A., and Sorbo, L. Can the universe afford inflation? *Physical Review D*, 70:063528, 2004.

Amrhein, V. Greenland, S., and McShane, B. Scientists rise up against statistical significance. *Nature*, 567:305–307, 2019.

Arkani-Hamed, N., Dimopoulos, S., and Dvali, G. The hierarchy problem and new dimensions at a millimeter. *Physics Letters B*, 429:263–272, 1998.

Arntzenius, F., and Dorr, C. Self-locating priors and cosmological measures. In K. Chamcham, J. Barrow, S. Saunders, and J. Silk, editors, *The Philosophy of Cosmology*, pages 396–428. Cambridge, UK: Cambridge University Press, 2017.

Azhar, F. Prediction and typicality in multiverse cosmology. *Classical and Quantum Gravity*, 31:035005, 2014.

Baker, D. J. Measurement outcomes and probability in Everettian quantum mechanics. *Studies in History and Philosophy of Modern Physics*, 38:153–169, 2007.

Barbieri, R., and Guidice, G. F. Upper bounds on suupersymmetric particle masses. *Nuclear Physics B*, 306:63–76, 1988.

Barnes, L. A. The fine-tuning of the universe for intelligent life. *Publications of the Astronomical Society of Australia*, 29:529–564, 2012.

Barnes, L. A. Fine-tuning in the context of Bayesian theory testing. *European Journal for Philosophy of Science*, 8:253–269, 2018.

Barnes, L. A., Elahi, P. J., Salcido, J. et al. Galaxy formation efficiency and the multiverse explanation of the cosmological constant with EAGLE simulations. *Monthly Notices of the Royal Astronomical Society*, 477:3727–3743, 2018.

Barr, S. M., and Khan, A. Anthropic Tuning of the weak scale and of m_u/m_d in two-Higgs-doublet models. *Physical Review D*, 76:045002, 2007.

Barrow, J. D., and Tipler, F. J. *The Anthropic Cosmological Principle*. Oxford: Oxford University Press, 1986.

Baun, L., and Frampton, P. H. Turnaround in cyclic cosmology. *Physical Review Letters*, 98:071301, 2007.

Behe, M. J. *Darwin's Black Box*. New York: The Free Press, 1996.

Bekenstein, J. D. Relativistic gravitation theory for the modified Newtonian dynamics paradigm. *Physical Review D*, 70:083509, 2004.

Bell, J. S. On the Einstein-Podolsky-Rosen paradox. *Physics*, 1:195–200, 1964.

Bell, J. S. On the problem of hidden variables in quantum mechanics. *Reviews of Modern Physics*, 38:447–452, 1966.

Bénétreau-Dupin, Y. Blurring out cosmic puzzles. *Philosophy of Science*, 82:879–891, 2015a.

Bénétreau-Dupin, Y. The Bayesian who knew too much. *Synthese*, 192:1527–1542, 2015b.

Bertone, G., and Tait, T. M. P. A new era in the search for dark matter. *Nature*, 562:51–56, 2018.

Bertone, G., Hooper, D., and Silk, J. Particle dark matter: Evidence, candidates and constraints. *Physics Reports*, 405:279–390, 2005.

Bertrand, J. L. F. *Calcul des probabilités*. Paris: Gauthier-Villars, 1889.

Boddy, K. K., Carroll, S. M., and Pollack, J. Why Boltzmann brains do not fluctuate into existence from the de Sitter vacuum. In K. Chamcham, J. Barrow, S. Saunders, and J. Silk, editors, *The Philosophy of Cosmology*, pages 228–240. Cambridge, UK: Cambridge University Press, 2017.

Bohm, D. A suggested interpretation of the quantum theory in terms of "hidden" variables, I and II. *Physical Review*, 85:166–193, 1952.

Boltzmann, L. On certain questions of the theory of gases. *Nature*, 51:413–415, 1895.

Borrelli, A., and Castellani, E. The practice of naturalness: A historical-philosophical perspective. *Foundations of Physics*, 49:860–878, 2019.

Bostrom, N. The doomsday argument, Adam & Eve, UN++, and Quantum Joe. *Synthese*, 127:359–387, 2001.

Bostrom, N. *Anthropic Bias: Observation Selection Effects in Science and Philosophy*. New York: Routledge, 2002.

Bostrom, N. Sleeping Beauty and self-location: A hybrid model. *Synthese*, 157:59–78, 2007.

Bostrom, N. Where are they? Why I hope the search for extraterrestrial life finds nothing. *MIT Technology Review*, May/June:72–77, 2008.

Bousso, R. Holographic properties in eternal inflation. *Physical Review Letters*, 97:191302, 2006.

Bousso, R. Complementarity in the multiverse. *Physical Review D*, 79:123524, 2009.

Bousso, R., Harnik, R., Kribs G. D., and Perez, G. Predicting the cosmological constant from the causal entropic principle. *Physical Review D*, 76:043513, 2007.

Bousso, R., Freivogel, B., and Yang, I. Boltzmann babies in the proper time measure. *Physical Review D*, 77:103514, 2008.

Bousso, R., Freivogel, B., and Yang, I. Properties of the scale factor measure. *Physical Review D*, 79:063513, 2009.

Bousso, R., and Polchinski, J. Quantization of four-form fluxes and dynamical neutralization of the cosmological constant. *Journal of High Energy Physics*, 2000:06, 2000.

Bradley, D. J. Multiple Universes and Observation Selection Effects. *American Philosophical Quarterly*, 46:61–72, 2009.

Bradley, D. J. Self-location is no problem for conditionalization. *Synthese*, 182:393–411, 2011.

Bradley, D. J. Four problems about self-locating belief. *Philosophical Review*, 121:149–177, 2012.

Bradley, D. J. Everettian confirmation and Sleeping Beauty: Reply to Wilson. *British Journal for the Philosophy of Science*, 66:683–693, 2015.

Bradley, D. J., and Leitgeb, H. When betting odds and credences come apart: More worries for Dutch book arguments. *Analysis*, 66:119–127, 2006.

Briggs, R. Putting a value on beauty. In T. S. Gendler and J. Hawthorne, editors, *Oxford Studies in Epistemology, Volume 3*, pages 3–34. Oxford: Oxford University Press, 2010.

Buckareff, A., and Nagasawa, Y., editors. *Alternative Concepts of God: Essays on the Metaphysics of the Divine*. Oxford: Oxford University Press, 2016.

Carlson, E., and Olsson, E. J. Is our existence in need of further explanation? *Inquiry*, 41:255–275, 1998.

Carr, B. J., and Rees, M. J. The anthropic principle and the structure of the physical world. *Nature*, 278:605–612, 1979.

Carretero-Sahuquillo, M. A. The charm quark as a naturalness success. *Studies in History and Philosophy of Modern Physics*, 58:51–61, 2019.

Carroll, S. M. *From Eternity to Here: The Quest for the Ultimate Theory of Time*. London: Plume, 2010.

Carroll, S. M. Why Boltzmann brains are bad. In S. Dasgupta and B. Weslake, editors, *Current Controversies in Philosophy of Science*. Routledge, 2020.

Carroll, S. M. Beyond falsifiability: Normal science in a multiverse. R. Dawid, R. Dardashti, and K. Thébault, editors, *Why Trust a Theory?* Cambridge, UK: Cambridge University Press, pages 300–314 of the book, 2019.

Carter, B. The anthropic principle and its implications for biological evolution. *Philosophical Transactions of the Royal Society of London*, A310:347–363, 1983.

Carter, B. J. Large number coincidences and the anthropic principle in cosmology. In M. S. Longair, editor, *Confrontation of Cosmological Theory with Astronomical Data*, pages 291–298. Dordrecht: Reidel, 1974.

Ćirković, M. M., Sandberg, A., and Bostrom, N. Anthropic shadow: Observation selection effects and human extinction risks. *Risk Analysis*, 30:1495–1506, 2010.

Clauser, J. F., Horne, M. A., Shimony, A., and Holt, R. A. Proposed experiment to test local hidden variables theories. *Physical Review Letters*, 23:880–884, 1969.

Colbeck, R., and Renner, R. No extension of quantum theory can have improved predictive power. *Nature Communications*, 2:411, 2011.

Collins, R. The teleological argument: An exploration of the fine-tuning of the cosmos. In W. L. Craig and J. P. Moreland, editors, *The Blackwell Companion to Natural Theology*, pages 202–281. Oxford: Blackwell, 2009.

Colyvan, M., Garfield, J. L., and Priest, G. Problems with the argument from fine-tuning. *Synthese*, 145:325–338, 2005.

Conitzer, V. A devastating example for the Halfer Rule. *Philosophical Studies*, 172:1985–1992, 2015a.

Conitzer, V. A Dutch Book against Sleeping Beauties who are evidential decision theorists. *Synthese*, 192:2887–2899, 2015b.

Corbelli, E., and Salucci, P. The extended rotation curve and the dark matter halo of M33. *Monthly Notices of the Royal Astronomical Society*, 311:441–447, 2000.

Cozic, M. Imaging and Sleeping Beauty: A case for double-halfers. *International Journal of Approximate Reasoning*, 52:137–143, 2011.

Craig, W. L. Design and the anthropic fine-tuning of the universe. In N. A. Manson, editor, *God and Design: The Teleological Argument and Modern Science*, pages 155–177. London: Routledge, 2003.

Curiel, E. Measure, topology and probabilistic reasoning in cosmology. 2014. latest version available online at http://philsci-archive.pitt.edu/11677/.

Davies, P. *The Goldilocks Enigma: Why Is the Universe Just Right for Life?* London: Allen Lane, 2006.

Dawid, R. *String Theory and the Scientific Method.* Cambridge, UK: Cambridge University Press, 2013.

Dawid, R., and Thébault, K. P. Against the empirical viability of the Deutsch-Wallace-Everett approach to quantum mechanics. *Studies in History and Philosophy of Modern Physics*, 47:55–61, 2014.

Dawid, R., and Thébault, K. P. Many worlds: Decoherent or incoherent. *Synthese*, 192:1559–1580, 2015.

Dawid, R. Hartmann, D., and Sprenger, J. The no alternatives argument. *British Journal for the Philosophy of Science*, 66:213–234, 2015.

Dawkins, R. *The Ancestor's Tale: A Pilgrimage to the Dawn of Evolution.* New York: Houghton Mifflin, 2004.

De Simone, A., Guth, A. H., Salem, M. P., and Vilenkin, A. Predicting the cosmological constant with the scale-factor cutoff measure. *Physical Review D*, 78:063520, 2008.

Dembski, W. A. *The Design Inference: Eliminating Chance through Small Probabilities.* Cambridge, UK: Cambridge University Press, 1998.

Deutsch, D. Quantum theory of probability and decisions. *Proceedings of the Royal Society of London. Series A. Mathematical, Physical and Engineering Sciences*, 455:3129–3137, 1999.

Dicke, R. H. Dirac's Cosmology and Mach's Principle. *Nature*, 192:440–441, 1961.

Dieks, D. Doomsday–Or: The dangers of statistics. *The Philosophical Quarterly*, 42:778–784, 1992.

Dieks, D. Reasoning about the future: Doom and beauty. *Synthese*, 156:427–439, 2007.

Dirac, P. A. M. A New Basis for Cosmology. *Proceedings of the Royal Society A*, 165:199–208, 1938.

Dizadji-Bahmani, F. The probability problem in Everettian quantum mechanics persists. *British Journal for the Philosophy of Science*, 66:257–283, 2015.

Donoghue, J. F. The fine-tuning problems of particle physics and anthropic mechanisms. In Bernard Carr, editor, *Universe of Multiverse?*, pages 231–246. Cambridge, UK: Cambridge University Press, 2007.

Dorr, C. Sleeping beauty: In defence of Elga. *Analysis*, 62:292–296, 2002.

Draper, K., and Pust, J. Probabilistic arguments for multiple universes. *Pacific Philosophical Quarterly*, 88:288–307, 2007.

Draper, K., and Pust, J. Diachronic dutch books and Sleeping Beauty. *Synthese*, 164:281–287, 2008.

Draper, P., Meade, P., Reece, M., and Shih, D. Implications of a 125 GeV Higgs boson for the MSSM and low-scale supersymmetry breaking. *Physical Review D*, 85:095007, 2012.

Earman, J. The SAP also rises: A critical examination of the anthropic principle. *Philosophical Quarterly*, 24:307–317, 1987.

Earman, J., and Mosterín, J. A critical look at inflationary cosmology. *Philosophy of Science*, 66:1–49, 1999.

Eckhardt, W. Probability theory and the Doomsday argument. *Mind*, 102:483–488, 1993.

Eddington, A. S. The end of the world: From the standpoint of mathematical physics. *Nature*, 127:447–453, 1931. reprinted in *The Book of the Cosmos: Imagining the Universe from Heraclitus to Hawking*, ed. by D. R. Danielson, Cambridge, MA: Perseus, 2000, p. 406.

Einstein, A. Autobiographical notes. In P. A. Schilpp, editor, *Albert Einstein: Philosopher-Scientist*, pages 1–94. Peru, IL: Open Court, 1949.

Elga, A. Self-locating belief and the sleeping beauty problem. *Analysis*, 60:143–147, 2000.

Elga, A. Defeating Dr. Evil with self-locating belief. *Philosophy and Phenomenological Research*, 69:383–396, 2004.

Ellis, G. F. R., and Stoeger, W. R. A note on infinities in eternal inflation. *General Relativity and Gravitation*, 41:1475–1484, 2009.

Epstein, P. F. The fine-tuning argument and the requirement of total evidence. *Philosophy of Science*, 84:639–658, 2017.

Everett, H. 'Relative state' formulation of quantum mechanics. *Reviews of Modern Physics*, 29:454–462, 1957.

Forrest, P. Occam's razor and possible worlds. *The Monist*, 65:456–464, 1982.

Forrest, P., and Armstrong, D. M. An argument against David Lewis' theory of possible worlds. *Australasian Journal of Philosophy*, 62:164–168, 1984.

Friederich, S. Motivating Wittgenstein's perspective on mathematical sentences as norms. *Philosophia Mathematica*, 19:1–19, 2011.

Friederich, S. *Interpreting Quantum Theory: A Therapeutic Approach*. Houndmills, Basingstoke: Palgrave Macmillan, 2014.

Friederich, S., and Evans, P. W. Retrocausality in quantum mechanics. In Edward N. Zalta, editor, *The Stanford Encyclopedia of Philosophy*. Metaphysics Research Lab, Stanford University, Summer 2019 edition, 2019. https://plato.stanford.edu/entries/qm-retrocausality/.

Fuchs, C. A., and Schack, R. QBism and the Greeks: Why a quantum state does not represent an element of physical reality. *Physica Scripta*, 90:015104, 2015.

Gaillard, M. K., and Lee, B. W. Rare decay modes of the K mesons in gauge theories. *Physical Review D*, 10:897, 1974.

Garber, D. Old evidence and logical omniscience in bayesian confimation theory. In J. Earman, editor, *Testing Scientific Theories*. Minneapolis: University of Minnesota Press, 1983.

Garriga, J. and Vilenkin, A. Prediction and explanation in the multiverse. *Physical Review D*, 77:043526, 2008.

Garriga, J., Schwartz-Perlov, D., Vilenkin, A., and Winitzki, S. Probabilities in the inflationary multiverse. *Journal of Cosmology and Astroparticle Physics*, 2006:017, 2006.

Ghirardi, G. C., Rimini, A., and Weber, T. Unified dynamics for microscopic and macroscopic systems. *Physical Review D*, 34:470, 1986.

Gibbons, G. W., Hawking, S. W., and Stewart, J. M. A natural measure on the set of all universes. *Nuclear Physics B*, 281:736–751, 1987.

Gibbons, G. W., and Turok, N. Measure problem in cosmology. *Physical Review D*, 77:063516, 2008.

Giudice, G. The dawn of the post-naturalness era. *CERN reports*, CERN-TH-2017-205, 2017. arXiv:1710.07663v1.

Glashow, S.L., Iliopoulos, J., and Maiani, L. Weak interactions with lepton-hadron symmetry. *Physical Review D*, 2:1285, 1970.

Glymour, C. *Theory and Evidence*. Princeton: Princeton University Press, 1980.

Goldstein, S., Struyve, W., and Tumulka, R. The Bohmian approach to the problem of cosmological quantum fluctuations. In A. Ijjas and B. Loewer, editors, *Guide to the Philosophy of Cosmology*. Oxford University Press, 2017. Preprint available at https://arxiv.org/abs/1508.01017.

Gould, S. J. Mind and supermind. *Natural History*, 92:34–38, 1983.

Greaves, H., and Myrvold, W. Everett and evidence. In S. Saunders, J. Barrett, A. Kent, and D. Wallace, editors, *Many Worlds? Everett, Quantum Theory and Reality*, pages 264–304. Oxford: Oxford University Press, 2010.

Greenberger, D. M., Horne, M. A., and Zeilinger, A. Going beyond Bell's theorem. In M. Kafatos, editor, *Bell's Theorem, Quantum Theory and Conceptions of the Universe*, pages 69–72. Dordrecht: Kluwer, 1989.

Greene, B. *The Hidden Reality*. New York: Vintage, 2011.

Grinbaum, A. Which fine-tuning arguments are fine? *Foundations of Physics*, 42:615–631, 2012.

Grohs, E., Howe, A. R., and Adams, F. C. Universes without the weak force: Astrophysical processes with stable neutrons. *Physical Review D*, 97:043003, 2018.

Guth, A., et al. A cosmic controversy. *Scientific American*, May 10, 2017. available online at https://blogs.scientificamerican.com/observations/a-cosmic-controversy/, accessed 23 August 2019.

Guth, A. H. Inflationary universe: A possible solution to the horizon and flatness problems. *Physical Review D*, 23:347–356, 1981.

Guth, A. H. Inflation and eternal inflation. *Physics Reports*, 333:555–574, 2000.

Guth, A. H. Eternal inflation and its implications. *Journal of Physics A*, 40:6811, 2007.

Hacking, I. The inverse gambler's fallacy: The argument from design; The anthropic principle applied to Wheeler Universes. *Mind*, 96:331–340, 1987.

Hall, L. J., Pinner, D., and Ruderman, J. T. The weak scale from BBN. *Journal of High Energy Physics*, 2014:134, 2014.

Halpern, J. Y. Sleeping Beauty reconsidered: Conditioning and reflection in asynchronoûs systems. In T. Gendler and J. Hawthorne, editors, *Oxford Studies in Epistemology*, pages 111–142. Oxford: Oxford University, 2005.

Harnik, R., Kribs, G. D., and Perez, G. A universe without weak interactions. *Physical Review D*, 2006:035006, 2006.

Hartle, J. B., and Srednicki, M. Are we typical? *Physical Review D*, 75:123523, 2007.

Hartmann, S., and Fitelson, B. A new Garber-style solution to the problem of old evidence. *Philosophy of Science*, 82:712–717, 2015.

Hawking, S. W., and Page, D. N. How probable is inflation? *Nuclear Physics B*, 298:789–809, 1988.

Hawthorne, J., and Isaacs, Y. Fine-tuning fine-tuning. In M. A. Benton, J. Hawthorne, and D. Rabinowitz, editors, *Knowledge, Belief, and God: New Insights in Religious Epistemology*, pages 136–168. Oxford: Oxford University Press, 2018.

Healey, R. A. *The Quantum Revolution in Philosophy*. Oxford: Oxford University Press, 2017.

Hill, C. T., and Simmons, E. H. Strong dynamics and electroweak symmetry breaking. *Physics Reports*, 381:235–402, 2003.

Hitchcock, C. Beauty and the bets. *Synthese*, 139:405–420, 2004.

Hogan, C. J. Why the universe is just so. *Reviews of Modern Physics*, 72:1149–1161, 2000.

Hogan, C. J. Quarks, electrons, and atoms in closely related universes. In Bernard Carr, editor, *Universe of Multiverse?*, pages 221–230. Cambridge, UK: Cambridge University Press, 2007.

Holder, R. D. Fine-tuning, multiple universes and theism. *Noûs*, 36:295–312, 2002.

Hollands, S., and Wald, R. M. Essay: An alternative to inflation. *General Relativity and Gravitation*, 34:2043–2055, 2002.

Horgan, T. Sleeping Beauty awakened: New odds at the dawn of the new day. *Analysis*, 64:10–21, 2004.

Hossenfelder, S. *Lost in Math: How Beauty Leads Physics Astray*. Basic Books, 2014.

Howson, C. The "old evidence" problem. *British Journal for the Philosophy of Science*, 42: 547–555, 1991.

Hoyle, F., Dunbar, D. N. F., Wenzel, W. A., and Whaling, W. A state in C12 predicted from astrophysical evidence. *Physical Review*, 92:1095, 1953.

Ijjas, A., Steinhardt, P. J., and Loeb, A. Inflationary paradigm in trouble after Planck2013. *Physics Letters B*, 723:547–555, 2013.

Ijjas, A., Steinhardt, P. J., and Loeb, A. Cosmic inflation theory faces challenges. *Scientific American*, February 1, 2017. Available online at www.scientificamerican.com/article/cosmic-inflation-theory-faces-challenges/, accessed 23 August 2019.

Jenkins, C. S. Sleeping Beauty: A wake-up call. *Philosophia Mathematica*, 13:194–201, 2005.

Juhl, C. Fine-tuning, many worlds, and the 'inverse gambler's fallacy'. *Noûs*, 39:337–347, 2005.

Juhl, C. Fine-tuning is not surprising. *Analysis*, 66:269–275, 2006.

Juhl, C. Fine-tuning and old evidence. *Noûs*, 41:550–558, 2007.

Kachru, S., Kallosh, R., Linde, A., and Trivedi, S. P. De Sitter vacua in string theory. *Physical Review D*, 68:046005, 2003.

Kent, A. One world versus many: The inadequacy of Everettian accounts of evolution, probability, and scientific confirmation. In S. Saunders, J. Barrett, A. Kent, and D. Wallace, editors, *Many Worlds? Everett, Quantum Theory and Reality*, pages 307–355. Oxford: Oxford University Press, 2010.

Keynes, J. M. *A Treatise on Probability*. London: Macmillan, 1921.

Kierland, B., and Monton, B. Minimizing inaccuracy for self-locating beliefs. *Philosophy and Phenomenological Research*, 70:384–395, 2005.

Knight, F. H. *Risk, Uncertainty, and Profit*. Boston: Hart, Schaffner & Marx, 1921.

Kochen, S., and Specker, E. P. The problem of hidden variables in quantum mechanics. *Journal of Mathematics and Mechanics*, 17:59–87, 1967.

Koperski, J. Should we care about fine-tuning? *British Journal for the Philosophy of Science*, 56:303–319, 2005.

Kotzen, M. Selection biases in likelihood arguments. *British Journal for the Philosophy of Science*, 63:825–839, 2012.

Kripke, S. Naming and Necessity. In D. Davidson and G. Harman, editors, *Semantics of Natural Language*, pages 253–355, 763–769. Dordrecht: Reidel, 1972. repr. 1980 by Harvard University Press.

Kroupa, P., Pawlowski, M., and Milgrom, M. The failures of the Standard Model of cosmology require a new paradigm. *International Journal of Modern Physics D*, 21: 1230003, 2012.

Landsman, K. The fine-tuning argument: Exploring the improbability of our own existence. In K. Landsman and E. van Wolde, editors, *The Challenge of Chance*, pages 111–128. Heidelberg: Springer, 2016.

Leegwater, G. An impossibility theorem for parameter independent hidden variable theories. *Studies in History and Philosophy of Modern Physics*, 54:18–34, 2016.

Lehners, J.-L., and Steinhard, P. J. Planck 2013 results support the cyclic universe. *Physical Review D*, 87:123533, 2013.

Lerche, W., Lüst, D., and Schellekens, A. N. Chiral four-dimensional heterotic strings from selfdual lattices. *Nuclear Physics B*, 287:477, 1987.

Leslie, J. Anthropic explanations in cosmology. In *PSA: Proceedings of the Biennial Meeting of the Philosophy of Science Association*, pages 87–95, 1986.

Leslie, J. No inverse gambler's fallacy in cosmology. *Mind*, 97:269–272, 1988.

Leslie, J. *Universes*. London: Routledge, 1989.

Lewis, D. Causal decision theory. *Australasion Journal of Philosophy*, 59:5–30, 1981.

Lewis, D. A subjectivists's guide to objective chance. In *Philosophical Papers, Vol. II*, pages 83–132. New York: Oxford University Press, 1986a. originally published in R. C. Jeffrey, editor, *Studies in Inductive Logic and Probability, Vol. II*, Berkeley: University of California Press, 1980.

Lewis, D. *On the Plurality of Worlds*. Oxford, New York: Blackwell, 1986b.

Lewis, D. Sleeping beauty: Reply to Elga. *Analysis*, 61:171–176, 2001.

Lewis, G. J., and Barnes, L. A. *Fortunate Universe: Life in a Finely Tuned Cosmos*. Cambridge, UK: Cambridge University Press, 2016.

Lewis, P. J. Quantum Sleeping Beauty. *Analysis*, 67:59–65, 2007.

Lewis, P. J. A note on the doomsday argument. *Analysis*, 70:27–30, 2010.

Linde, A. A new inflationary universe scenario: A possible solution of the horizon, flatness, homogeneity, isotropy and primordial monopole problems. *Physics Letters B*, 108:389–393, 1982.

Linde, A. Sinks in the landscape, Boltzmann brains and the cosmological constant. *Journal of Cosmology and Astroparticle Physics*, 2007:002, 2007.

Linde, A. Inflationary cosmolgoy after Planck 2013. 2014. Available online at arXiv:1402.0526, retrieved 3 January 2019.

Linde, A., Linde, D., and Mezhlumian, A. Nonperturbative amplifications of inhomogeneities in a self-reproducing universe. *Physical Review D*, 54:2504, 1995.

Loeb, A., Batista, R. A., and Sloan, D. Relative likelihood for life as a function of cosmic time. *Journal of Cosmology and Astroparticle Physics*, 08:40, 2016.

MacDonald, J., and Mullan, D. J. Big bang nucleosynthesis: The strong nuclear force meets the weak anthropic principle. *Physical Review D*, 80:043507, 2009.

Manson, N. A. The fine-tuning argument. *Philosophy Compass*, 492:29, 2009.

Manson, N. A. How not to be generous to fine-tuning sceptics. *Religious Studies*, 2018. https://doi.org/10.1017/S0034412518000586.

Manson, N. A., and Thrush, M. J. Fine-tuning, multiple universes, and the "This Universe" Objection. *Pacific Philosophical Quarterly*, 84:67–83, 2003.

Martel, H., Shapiro, P. R., and Weinberg, S. Likely values of the cosmological constant. *The Astrophysical Journal*, 492:29, 1998.

Martin, J. Cosmic inflation: Trick or treat? In *Fine-tuning in the Physical Universe*. Cambridge, UK: Cambridge University Press, in press arXiv:1902.02586v1.

McCoy, C. D. Does inflation solve the hot big bang model's fine-tuning problems? *Studies in History and Philosophy of Modern Physics*, 51:23–36, 2015.

McCoy, C. D. The implementation, interpretation, and justification of likelihoods in cosmology. *Studies in History and Philosophy of Modern Physics*, 62:19–35, 2018.

McGrath, P. J. The inverse gambler's Fallacy and cosmology: A reply to hacking. *Mind*, 97:331–340, 1988.

McGrew, T., McGrew, L., and Vestrup, E. Probabilities and the fine-tuning argument: A sceptical view. *Mind*, 110:1027–1038, 2001.

McKenzie, K. Arguing against fundamentality. *Studies in History and Philosophy of Modern Physics*, 42:244–255, 2011.

McMullin, E. Indifference principle and anthropic principle in cosmology. *Studies in History and Philosophy of Science*, 24:359–389, 1993.

McQueen, K. J., and Vaidman, L. In defence of the self-location uncertainty account of probability in the many-worlds interpretation. *Studies in History and Philosophy of Modern Physics*, 66:14–23, 2019.

Meacham, C. J. G. Sleeping Beauty and the dynamics of *de se* belief. *Philosophical Studies*, 138:245–2699, 2008.

Milgrom, M. A modification of the Newtonian dynamics as a possible alternative to the hidden mass hypothesis. *Astrophysical Journal*, 270:365–370, 1983.

Miller, K. R., *Finding Darwin's God: A Scientist's Search for Common Ground between God and Evolution*. New York: Cliff Street Books, 1999.

Monton, B. God, fine-tuning, and the problem of old evidence. *British Journal for the Philosophy of Science*, 57:404–425, 2006.

Mukhanov, V. Inflation without selfreproduction. *Fortschritte der Physik*, 63:36–41, 2015.

Narveson, J. God by design? In *God and Design: The Teleological Argument and Modern Science*, pages 88–105. London: Routledge, 2003.

Neal, R. M. Puzzles of anthropic reasoning resolved using full non-indexical conditioning. 2006. arXiv:math/0608592v1.

Norton, J. D. Ignorance and indifference. *Philosophy of Science*, 75:45–68, 2008.

Norton, J. D. Cosmic confusion: Not supporting versus supporting not. *Philosophy of Science*, 77:501–523, 2010.

Norton, J. D. Eternal inflation: When probabilities fail. *Synthese*, in press https://doi.org/10.1007/s11229-018-1734-7.

Oberhummer, H., Csótó, A., and Schlattl, H. Stellar production rates of carbon and its abundance in the universe. *Science*, 289:88–90, 2000.

Olum, K. The Doomsday Argument and the number of possible observers. *The Philosophical Quarterly*, 52:164–184, 2002.

Page, D. N. Is our universe likely to decay within 20 billion years? *Physical Review D*, 78:063535, 2008.

Parfit, D. Why Anything? Why This?. *London Review of Books*, January 22:24–27, 1998.

Pearl, J. *Causality*. New York: Cambridge University Press, 2000.

Penrose, R. *The Road to Reality: A Complete Guide to the Laws of the Universe*. London: Vintage, 2004.

Phillips, D., and Albrecht, A. Effects of inhomogeneity on the causal entropic prediction of Λ. *Physical Review D*, 84:123530, 2011.

Pisaturo, R. Past longevity as evidence for the future. *Philosophy of Science*, 76:73–100, 2009.

Planck Collaboration. Planck 2015 results. I. Overview of products and scientific results. *Astronomy and Astrophysics*, 594:A1, 2016.

Price, H. Against causal decision theory. *Synthese*, 67:195–212, 1986.

Randall, L., and Sundrum, R. Large mass hierarchy from a small extra dimension. *Physical Review Letters*, 83:3370, 1999.

Rees, M. *Just Six Numbers: The Deep Forces that Shape the Universe*. New York: Basic Books, 2000.

Rees, M. *Our Final Hour: A Scientist's Warning*. New York: Basic Books, 2003.

Roberts, J. T. Fine-tuning and the infrared bull's eye. *Philosophical Studies*, 160:287–303, 2012.

Rosaler, J., and Harlander, R. Naturalness, Wilsonian renormalization, and "fundamental parameters" in quantum field theory. *Studies in History and Philosophy of Modern Physics*, 66:118–134, 2019.

Ross, J. Sleeping Beauty, countable additivity, and rational dilemmas. *Philosophical Review*, 119:411–447, 2010.

Rota, M. *Taking Pascal's Wager: Faith, Evidence, and the Abundant Life*. Downers Grove, IL: Intervarsity Press, 2016.

Salem, M. P. Bubble collisions and measures of the multiverse. *Journal of Cosmology and Astroparticle Physics*, 2012(01):021, 2012.

Sandberg, A., Drexler, E., and Ord, T. Dissolving the Fermi paradox. 2018. arXiv:1806.02404.

Saunders, S. Derivation of the Born rule from operational assumptions. *Proceedings of the Royal Society of London. Series A: Mathematical, Physical and Engineering Sciences*, 460:1771–1788, 2004.

Schellekens, A. N. Life at the interface of particle physics and string theory. *Reviews of Modern Physics*, 85:1491, 2013.

Schulz, M. The dynamics of indexical belief. *Erkenntnis*, 72:337–351, 2010.

Schwarz, W. Belief update across fission. *British Journal for the Philosophy of Science*, 66:659–682, 2015.

Sebens, C. T., and Carroll, S. M. Self-locating uncertainty and the origin of probability in Everettian quantum mechanics. *British Journal for the Philosophy of Science*, 2018 (69):25–74, 2018.

Shackel, N. Bertrand's paradox and the principle of indifference. *Philosophy of Science*, 74:150–175, 2007.

Shiffrin, J. S., and Wald, R. M. Measure and probability in cosmology. *Physical Review D*, 86:023521, 2012.

Simmons, J. P., Nelson, L. D., and Simonsohn, U. False-positive psychology: Undisclosed flexibility in data collection and analysis allows presenting anything as significant. *Psychological Science*, 22:1359–1366, 2011.

Smart, J. J. C. *Our Place in the Universe: A Metaphysical Discussion*. Oxford: Blackwell, 1989.

Smeenk, C. Predictability crisis in early universe cosmology. *Studies in History and Philosophy of Modern Physics*, 46:122–133, 2014.

Smolin, L. *The Trouble with Physics: The Rise of String Theory, The Fall of a Science, and What Comes Next*. New York: Houghton Mifflin, 2006.

Smolin, L. Scientific alternatives to the anthropic principle. In B. Carr, editor, *Universe of Multiverse*, pages 323–366. Cambridge, UK: Cambridge University Press, 2007.

Sober, E. The design argument. In N. A. Manson, editor, *God and Design: The Teleological Argument and Modern Science*, pages 27–54. London: Routledge, 2003.

Sober, E. Absence of evidence and evidence of absence: Evidential transitivity in connection with fossils, fishing, fine-tuning and firing squads. *Philosophical Studies*, 143:63–90, 2009.

Spekkens, R. W Contextuality for preparations, transformations, and unsharp measurements. *Physical Review A*, 71:052108, 2005.

Sprenger, J. A novel solution to the problem of old evidence. *Philosophy of Science*, pages 383–401, 2015.

Srednicki, M., and Hartle, J.B. Science in a very large universe. *Physical Review D*, 81:123524, 2010.

Starkman, G. D., and Trotta, R. Why anthropic reasoning cannot predict Λ. *Physical Review Letters*, 97:201301, 2006.

Steinhardt, P. J. The inflation debate. *Scientific American*, April:36–43, 2011.

Steinhardt, P. J., and Turok, N. Cosmic evolution in a cyclic universe. *Physical Review D*, 65:126003, 2001.

Steinhardt, P. J. and Turok, N. *Endless Universe: Beyond the Big Bang*. New York: Doubleday, 2008.

Stenger, V. J. *The Fallacy of Fine-tuning: Why the Universe Is Not Designed for Us*. New York: Prometheus Books, 2011.

Susskind, L. *The Cosmic Landscape: String Theory and the Illusion of Intelligent Design*. New York: Back Bay Books, 2005.

Swimburne, R. The argument to God from fine-tuning reassessed. In N. A. Manson, editor, *God and Design: The Teleological Argument and Modern Science*, pages 105–123. London: Routledge, 2003.

Swimburne, R. *The Existence of God*. Oxford: Oxford University Press, 2nd edition, 2004.

Swimburne, R. *The Coherence of Theism*. Oxford: Oxford University Press, 2nd edition, 2016.

't Hooft, G. Naturalness, chiral symmetry and spontaneous chiral symmetry breaking. In G. 't Hooft , editor, *Recent Developments in Gauge Theories*, pages 135–157. New York: Plenum Press, 1980.

Tegmark, M. Is "the theory of everything" merely the ultimate ensemble theory? *Annals of Physics*, 270:1–51, 1998.

Tegmark, M. What does inflation really predict? *Journal of Cosmology and Astroparticle Physics*, 2005:001, 2005.

Tegmark, M. *Our Mathematical Universe: My Quest for the Ultimate Nature of Reality*. New York: Knopf, 2014.

Tegmark, M. Aguirre, A. Rees, M. J., and Wilczek, F. Dimensionless constants, cosmology, and other dark matters. *Physical Review D*, 73:023505, 2006.

Tegmark, M., and Rees, M. J. Why is the cosmic microwave background fluctuation level 10^{-5}? *The Astrophysical Journal*, 499:526–532, 1998.

Thomas, H. Modal realism and inductive scepticism. *Noûs*, 27:331–354, 1993.

Titelbaum, M. G. The relevance of self-locating beliefs. *Philosophical Review*, 117:555–605, 2008.

Titelbaum, M. G. An embarrassment for double-halfers. *Thought*, 1:146–151, 2012.

Titelbaum, M. G. *Quitting Certainties: A Bayesian Framework Modelling Degrees of Belief*. Oxford: Clarendon, revised edition, 2013a.

Titelbaum, M. G. Ten reasons to care about the Sleeping Beauty problem. *Philosophy Compass*, 8:1003–1017, 2013b.

Tolman, R. C. *Relativity, Thermodynamics, and Cosmology*. Oxford: Clarendon, 1934. reissued 1987 by Dover, New York.

Torres, P. *Morality, Foresight, and Human Flourishing: An Introduction to Existential Risks*. Durham, NC: Pitchstone, 2017.

Uzan, J.-Ph. The fundamental constants and their variation: Observational and theoretical status. *Reviews of Modern Physics*, 75:403, 2003.

Uzan, J.-Ph. Varying constants, gravitation and cosmology. *Living Reviews in Relativity*, 14:2, 2011.

Vaidman, L. On schizophrenic experiences of the neutron or why we should believe in the many-worlds interpretation of quantum theory. *International Studies in the Philosophy of Science*, 12:245–261, 1998.

van Fraassen, B. *Laws and Symmetry*. Oxford: Clarendon, 1989.

van Inwagen, P. *Metaphysics*. Colorado: Westview Press, 1993.

Van Schaik, C., and Michel, K. *The Good Book of Human Nature: An Evolutionary Reading of the Bible*. New York: Basic Books, 2016.

Venn, J. *The Logic of Chance*. New York: Chelsea, 1866.

Vilenkin, A. Predictions from quantum cosmology. *Physical Review Letters*, 74:846–849, 1995.

Vilenkin, A. A measure of the multiverse. *Journal of Physics A*, 40:6777–6785, 2007.

Wallace, D. Quantum probability from subjective likelihood: Improving on Deutsch's proof of the probability rule. *Studies in History and Philosophy of Modern Physics*, 38:311–332, 2007.

Wallace, D. *The Emergent Multiverse: Quantum Theory according to the Everett Interpretation*. Oxford: Oxford University Press, 2012.

Ward, P., and Brownlee, D. E. *Rare Earth: Why Complex Life Is Uncommon in the Universe*. New York: Copernicus, 2000.

Weatherson, B. Should we respond to evil with indifference? *Philosophy and Phenomenological Research*, 70:613–635, 2005.

Weinberg, S. Anthropic bound on the cosmological constant. *Physical Review Letters*, 59:2607, 1987.

Weinstein, S. Anthropic reasoning and typicality in multiverse cosmology and string theory. *Classical and Quantum Gravity*, 23:4231, 2006.

Weisberg, J. Firing squads and fine-tuning: Sober on the design argument. *British Journal for the Philosophy of Science*, 56:809–821, 2005.

Weisberg, J. A note on design: What's fine-tuning go to do with it? *Analysis*, 70:431–438, 2010.

Weisberg, J. The argument from divine indifference. *Analysis*, 72:707–715, 2012.

Wells, J. D. The utility of Naturalness, and how its application to Quantum Electrodynamics envisages the Standard Model and Higgs boson. *Studies in History and Philosophy of Modern Physics*, 49:102–108, 2015.

Wetterich, C. Fine-tuning problem and the renormalization group. *Physics Letters B*, 140:215–222, 1984.

Wetterich, C. Where to look for solving the gauge hierarchy problem? *Physics Letters B*, 718:573–576, 2012.

White, R. Fine-tuning and multiple universes. *Noûs*, 34:260–267, 2000.

White, R. What's fine-tuning got to do with it: A reply to Weisberg. *Analysis*, 71:676–679, 2011.

Williams, P. Naturalness, the autonomy of scales, and the 125GeV Higgs. *Studies in History and Philosophy of Modern Physics*, 51:82–96, 2015.

Wilson, A. Everettian confirmation and Sleeping Beauty. *British Journal for the Philosophy of Science*, 65:573–598, 2014.

Woit, P. *Not Even Wrong: The Failure of String Theory and the Search for Unity in Physical Law*. Basic Books, 2006.

Woodward, J. *Making Things Happen: A Theory of Causal Explanation*. Oxford: Oxford University Press, 2003.

Zuboff, A. One self: The logic of experience. *Inquiry*, 33:39–68, 1990.

Zurek, W. H. Probabilities from entanglement, Born's rule $p_k = |\psi_k|^2$ from envariance. *Physical Review A*, 71:052105, 2005.

Author Index

Subject Index